The Quality of Measurements

T0137445

A.E. Fridman

The Quality of Measurements

A Metrological Reference

Translated by Andrew Sabak and Paul Makinen

 Springer

A.E. Fridman
Saint Petersburg State
Polytechnic Institute
Russia

ISBN 978-1-4899-9975-7   ISBN 978-1-4614-1478-0 (eBook)
DOI 10.1007/978-1-4614-1478-0
Springer New York Dordrecht Heidelberg London

Printed on acid-free paper

Springer is part of Springer Science+Business Media (www.springer.com)

Foreword

February 8, 2009 is the 175th anniversary of the birth of Dmitrii Ivanovich Mendeleev, the great Russian scientist and statesman. All of his wide-ranging scientific activities were oriented towards strengthening and developing the Russian state: science, education, industry, and trade. In his notebook, he wrote: "I . . . have endeavored, and will endeavor, as long as I am able, to contribute to the country in a fruitful, industrially realistic way, confident that the country's politics, development, education, and even defense is now unthinkable without development of industry. . .Science and industry – those are my dreams."

D.I. Mendeleev made a fundamental contribution to the development of metrology, both domestically and around the world. In 1892, he headed the first government metrological institution in Russia – the Depositary of Standard Weights and Measures, transforming it into a scientific research center of world significance – the Main Chamber of Weights and Measures (currently known as the D.I. Mendeleev All-Russian Scientific Research Institute for Metrology). Mendeleev's genius was fully apparent in that even in the early twentieth century he understood the nature of the internal relationship between metrology and the level of scientific and industrial development. His metrological reform marked for the first time that an infrastructure for ensuring uniformity of measurements had been established under the auspices of the Russian government; the scientific, legal, and organizational bases of this infrastructure have continued to remain important, even into the present day. D.I. Mendeleev developed a general approach to metrological research and encouraged domestic metrological research to be performed using his own trademarked personal style, which involved a preliminary deep study of the item under study, extremely careful formulation of the experiment, a detailed study of all sources of error, and mandatory reduction of the research results to practice. He founded the Russian metrological school, and initiated professional training of metrologists. Mendeleev's traditions have been carefully guarded by several generations of VNIIM scientific personnel, who can rightfully be considered students and disciples of the great metrologist.

One of Mendeleev's legacies has been the special attention paid to popularization (in light of the "great impact of properly directed education on the success of the country") and, in particular, improving the quality of metrological education in Russia. This metrology book was written for the great scientist's anniversary by one of the oldest staff members of the D.I. Mendeleev All-Russian Scientific Research Institute for Metrology, Dr. (Eng. Sci.) A.E. Fridman.

St. Petersburg, Russia   N.I. Khanov

From the Author

I was prompted to write this book by my experience in teaching metrology at the Saint Petersburg State Polytechnic Institute (SPbGPU). Metrology is a unique scientific discipline that occupies an intermediate position between the fundamental and applied sciences. The study of metrology is extremely valuable in the training of scientists, primarily because it assists in forming the scientific worldview of young specialists and accustoms them to thinking from a metrological point of view, and to thinking from the point of view of a stochastic approach to quantitative data on objects and phenomena in the tangible world.

Moreover, the role played by measurements in the modern world makes it essential to study metrology. Measurements are an important part of many areas in modern high technology: industrial production, scientific research, medicine, defense, and many others. Scientists and technical specialists must therefore have the theoretical knowledge and practical skills to develop measurement procedures, perform measurements, and assess and interpret the results of measurements, i.e., the knowledge and skills imparted by a course in metrology.

These goals could be more readily achieved by including a mandatory metrology course in the curricula for all technical specialties at the university level. However, none of the textbooks and books that I know of in this field meet the current requirements since they do not reflect the enormous changes that have occurred in metrology in recent years. This statement refers not only to the textbooks that have been prominent fixtures on the desks of several generations of Russian metrologists – the book by M.F. Malikov [1], textbooks [2, 3], and several later books [4–7, etc.].

The changes referred to above have affected virtually all of the major areas of metrological activity. Several radically new approaches to metrology have been theoretically validated and implemented. For example, a new system for the assessment of measurement accuracy based on the concept of measurement uncertainty has become generally accepted, even becoming an international regulation. A radically new system for assurance of measurement accuracy has seen wider and wider use both in this country and around the world: This system uses measures of convergence and reproducibility (obtained by comparison of results obtained from measurements of the same quantity by a group of measurement laboratories) in place of measures of measurement error (obtained through calibration of measurement instruments against standards).

The traditional systems for realization and dissemination of units, which served throughout the twentieth century as the basis for ensuring traceability of measurements, can no longer serve that purpose, now that traceability of measurements is required on a worldwide level. Therefore, these systems, which function as before on the national level, have been supplemented by an international system for confirming metrological equivalence of national standards based on key comparisons and regional comparisons under the auspices of international metrological organizations.

These changes have significantly affected the content of metrological activities in Russia and other countries around the world. At the same time, implementation of these new ideas should not lead to a rejection of the classical approaches, which continue to play a key role in practical metrology. Making the new and classical approaches consistent has had a substantial impact on the system of metrological terms and concepts; this impact has included the development of several new concepts, revision of our understanding of many existing concepts, and a reformulation of these concepts.

All of these scientific results (and many others) that have been obtained over the past decade and that have seen widespread use in practical metrology are covered in this book, which is an expanded version of the lecture notes for a metrology course I taught at the Saint Petersburg State Technical Institute.

Chapter 1 discusses the basic concepts of metrology from a modern point of view. Chapter 2 is devoted to measurement errors. For the readers' convenience, the book includes the minimum information required from probability theory and statistics. The theory of random measurement errors, which shows that these errors follow a generalized normal distribution law (and, in the most common special case – a normal distribution law), is then discussed. A mathematical description of instrumental systematic error is provided as a function of the metrological characteristics of the measurement system and as a function of the measurement conditions. Procedural measurement error is also discussed, along with techniques for eliminating systematic error. Chapter 3 discusses the concept of measurement uncertainty, which is now covered by international regulation [19]. This concept is shown to be compatible and consistent with the classical theory of measurement errors. The method for determining measured results (and for estimating the uncertainty in said results) implemented by this regulation is also described. Chapter 4 provides a detailed discussion of methods for statistical reduction of measurement results (revised in light of this procedure).

Chapter 5 describes the basic assumptions and principles underlying the design of the International System of Units (SI). Chapter 6 is devoted to traceability of measurements. A detailed discussion of the major elements of national systems for traceability of measurements is provided: references, calibration schemes, and organizational structures. The metrological and legal characteristics of the two forms for dissemination of units – calibration and verification – are described; a mathematical description of the statistical errors in verification is provided; and a mechanism for determining the value of the uncertainty in the magnitude of a unit during such operations will be provided. A new international system for assuring traceability of measurements is described; this system is based on an agreement among the directors of the various national metrological institutes, as well as key comparisons among national standards. Chapter 7 describes methods for determination and adjustment of the verification and calibration periods. Since these methods are grounded in the theory underlying the metrological reliability of the SI system, they will therefore be discussed simultaneously with a presentation of the basic provisions of this theory. Chapter 8 describes a new methodology for assurance of measurement accuracy (based on the provisions contained in the ISO 5725-series standards) that is now becoming more and more common around the world.

Since this book is primarily addressed to metrology students and instructors, all metrological quantities defined in the text are in italics.

A.E. Fridman

Contents

Chapter 1
Basic Concepts in Metrology

1.1 Introduction

1.2 Properties and Quantities

It will be convenient to begin our discussion of metrological concepts with the concepts of property and magnitude. *A property of an object is one of its distinguishing traits or characteristics. Heavy, long, strong, bright – these are all examples of properties of various objects. In philosophy, property is defined as a philosophical category, which expresses that aspect of an object that determines whether it has something in common with other objects or is different from other objects. Properties are qualitative characteristics, and many of them cannot be expressed quantitatively. Other properties can be expressed quantitatively. Such properties are called quantities. A quantity is a property that many objects (states, systems, and processes) have in common on a qualitative basis, but the quantitative value of the property for each object is specific to that object.*

An extremely general classification of quantities as a concept is shown in Fig. 1.

Quantities are initially divided into real quantities and ideal quantities. An ideal quantity is any numerical value. It is, by its very essence, an abstraction, not associated with any real object. Therefore, ideal quantities are studied in mathematics rather than metrology.

Real quantities are divided into physical quantities and ideal quantities. Non-physical quantities are introduced, determined, and studied in information science, the social sciences, economic sciences, and humanitarian sciences (e.g., sociology or linguistics). Examples of non-physical quantities are as follows: Amount of information in bits, amount of financial capital in dollars, and a variety of ratings determined using sociological surveys. The physical quantities that

A.E. Fridman, *The Quality of Measurements: A Metrological Reference*,
DOI 10.1007/978-1-4614-1478-0_1, © Springer Science+Business Media, LLC 2012

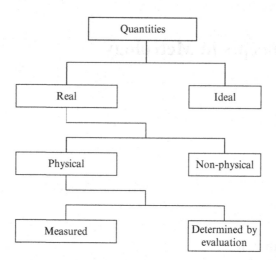

Fig. 1 Classification of quantities

metrology deals with are the properties of tangible objects, processes, and phenomena. Unlike non-physical quantities, they exist in the tangible world around us in an objective manner, independent of human desires.

Physical quantities are divided with respect to the method used for quantitative determination into measured quantities and estimated quantities. The distinguishing feature of a measurement is the presence of a measuring instrument – a special piece of equipment that stores the magnitude of a unit and is used to determine the value of a quantity. The process of estimation primarily refers to expert assessments and organoleptic assessments (i.e., assessments using human sensory organs) of quantities, such as, for example, determination of distances "by eye." In such cases, there is no piece of equipment in which to store the magnitude of a unit, meaning that there is no assurance that the estimate obtained meets accuracy requirements. The magnitude of the standard length dimension maintained by a person in his consciousness varies substantially from one individual to another, and this value can also vary for a single individual depending on his psychological and physical condition. Thus, such an estimate is inaccurate and unreliable, and there is no guarantee that the result of the estimation process will be objective. Only a piece of equipment not subject to human deficiencies is capable of providing such a guarantee. This is in fact the reason that measurements are the highest form of quantitative estimation.

The history of metrology indicates that all physical quantities follow what is essentially an identical path. Following discovery and identification of a new property and determination of a physical quantity: first, a method for quantitative estimation of the quantity is developed, and then, as knowledge increases, estimation is replaced by indirect measurements. Measures and approaches for direct measurements of this quantity will then be developed and used as a basis for establishment of a system for metrological traceability of the new form of measurements. For example, this was the path followed by measurements of color: from color atlases to a form of measurement – colorimetry, including equipment and measurement techniques, as well as metrologic traceability of said measurements. Acoustic measurements and salinity measurements followed a similar path. Measurement techniques will obviously be developed for many other physical quantities that are currently estimated via expert or organoleptic methods, such that said quantities become measured quantities.

The above analysis enables us to draw a clear boundary between measured and estimated quantities: Measured quantities are physical quantities for which measurement techniques already exist, and estimated quantities are physical quantities for which measurement techniques have yet to be developed.

1.3   Magnitude and Value of a Quantity. Units of Measurement for Quantities. Reference Scales for Quantities

In actual fact, it is not the quantities themselves, but the quantities associated with specific objects, phenomena, or processes, i.e., quantities with specific dimensions, that are measured. Strictly speaking, *measurement is the act of estimating the magnitude of a quantity using specialized equipment* (*). The *magnitude of a quantity is understood to be a quantitative determination of a quantity related to a specific tangible object, system, phenomenon, or process.*

Upon comparing the magnitude of some quantity to the unit for that quantity, we obtain the *value* of the quantity. Thus, the *value of a quantity expresses the magnitude of a quantity in the form of some number of the units adopted for that quantity.* In formal notation, this is written as follows:

$$X = x \times [X],\qquad\qquad(1.1)$$

where X is the value of the quantity,

[X] is the unit of measurement for the quantity,

x is an abstract number occurring in the value of the quantity. It is called the numerical value of the quantity.

Equation (1.1) is called the *fundamental equation of measurement*.

Setting $x = 1$ in (1.1), we find that: $X = 1 \times [X]$. This then implies the following definition of a unit:

A unit of measure (or, briefly, a unit) is understood to be a quantity of specific magnitude that is arbitrarily assigned a numerical value of 1 and is used for quantitative expression of homogeneous quantities. For example, the unit of length is 1 m (m) and the unit of mass is 1 kg (kg).

Upon comparing (1.1) with definition (*), the following definition of the term "measurement" can be provided:

Measurement is the act of comparing a physical quantity against the unit for that quantity using specialized equipment (**).

However, this definition does not include many physical quantities that are measured using equipment, but for which units are not available due to considerations of a fundamental nature. On the one hand, since no units are available, this is considered "estimation" rather than "measurement" under definition (**). On the other hand, if the problem is posed in this manner, many equipment-determined physical quantities of manufacturing or social importance might turn out to be beyond the scope of practical metrology, which would, of course, be unacceptable. In order to resolve this contradiction, definition (**) needs to be expanded. This is done using the concept of a *quantity-value scale*. By analogy with the definition of "unit," a "quantity-value scale" can be defined as follows:

A "quantity-value scale" is defined as an ordered set of quantity values (manifestations of properties) that are arbitrarily assigned to certain specific values used for quantitative expression of quantities (properties) that are homogeneous with said ordered set. Five basic types of scales are distinguished based on the logical structure behind manifestation of the properties [8]:

1. Scale of names (scale of classifications). Scales of this type are used for classification of objects whose properties only manifest themselves in the form of equivalence or non-equivalence. The only statement that can be made with respect to two objects A and B within this class is the following: $A = B$ (A is identical to B) or $A \neq B$. These properties cannot be described quantitatively, and the scales for these properties do not include the concepts of "larger," "smaller," "zero," or "one." One example of this type of scale would be classification of object colors based on the names (red, green, etc.) contained in standard color atlases. Other examples of name scales include classifications of components based on appearance/type, power consumption, etc. $A \ B$

2. *Order scale (rank scale)* This scale is monotonically increasing or decreasing, and enables establishment of equivalence and order relationships ($A<B$ or $A>B$) among quantities describing this property. In an order scale, it is possible to establish a hierarchy of objects with respect to the property being assessed, i.e., if $A<B$ and $B<C$, then $A<C$. The concept of "order scale" includes arbitrary scales – scales whose values are given in arbitrary units (for example, the 12-value Beaufort marine wind scale). Order scales and the reference points on order scales have become very common. The various hardness scales – Brinell, Rockwell, Vickers, etc. – are classified as order scales. For example, the Mohs scale for determination of mineral hardness contains ten reference minerals that are assigned arbitrary hardness numbers: talc–1, gypsum–2, calcium–3, fluorite–4, apatite–5, orthoclase–6, quartz–7, topaz–8, corundum–9, and diamond–10. Mineral hardnesses are estimated by scratching with the reference minerals. For example, if scratching the mineral with quartz (7) leaves a mark, and orthoclase (6) does not, the hardness of this mineral satisfies the inequality $6<Q<7$.

3. *Interval scale (difference scale).* These scales are used for objects in which the differences among properties satisfy equivalence, order, and additivity relations (an additive quantity is a homogeneous physical quantity whose values may be added, multiplied by a numerical factor, divided by one another; thus, if $A - B = q_1$ and $B - C = q_2$, then $A - C = (A - B) + (B - C) = q_1 + q_2$, $(A - B)p = q_1 p$, and $(A - B)/(B - C) = q_1/q_2$. An interval scale consists of identical intervals, has a "1" and one arbitrarily chosen zero point. The Celsius, Fahrenheit, and Reaumur temperature scales are interval scales, as is the count of years in various calendars.

   Interval scales have arbitrary (adopted by agreement) units and arbitrary zeros based on some sort of reference. By analogy with (1), they are described using the equation

$$Q = Q_0 + q[Q] \tag{1.2}$$

   where Q is the value of the quantity, $[Q]$ is the unit of the quantity, q is a numerical value, and Q_0 is the zero point of the scale. We see that the interval between any two quantities Q_1 and Q_2, which is equal to $Q_1 - Q_2 = (q_1 - q_2) \times [Q]$ satisfies the fundamental equation of measurement (1.1). Thus, on such scales, the difference between two quantities is also a quantity that has a specific physical meaning. However, the sum of quantities $Q_1 + Q_2 = 2Q_0 + (q_1 + q_2)[Q]$ satisfies neither the fundamental equation of measurement (1.1) nor the equation for the interval scale (1.2). This means that the sum of quantities does not have any physical meaning on an interval scale. For example, the difference in temperature between two bodies is the temperature difference by which one of the bodies must be heated in order for their temperatures to become equal, while the sum of two temperatures has no physical meaning. In exactly the same way, the difference between two dates on the calendar has clear physical meaning as the duration of the period of time between these two dates, while the sum of the dates has no physical meaning.

4. *Ratio scale.* These scales describe object properties that satisfy equivalence, order, and additivity relationships. Mass scales, length scales, and current scales (and many other scales) are examples of such scales. Ratio scales have arbitrary units and natural zeros. They therefore satisfy the equation

$$Q = q[Q]. \tag{1.3}$$

   The fact that the sum of the quantities $Q_1 = q_1[Q]$ and $Q_2 = q_2[Q]$, which is equal to $(Q_1 + Q_2) = (q_1 + q_2)[Q]$, satisfies fundamental equation of measurement (1.1) implies that both differences of quantities and sums of quantities have physical meaning in ratio scales.

5. *Absolute scale* When measuring relative quantities (ratios of homogeneous quantities: molar concentration of a component, power factor, coefficient of friction, and many others), the unit of measure is introduced in a natural way: $[Q] = 1$. In this case, the scale equation takes the form $Q = q$. Such scales, which

are also ratio scales, are frequently called "absolute scales." Dimensionless and count units of absolute scales are used for generation of many SI derived units.

Using the above information on scales of quantities, we can now revise definition (**):

Measurement is the act of comparing a physical quantity against the unit or scale for that quantity using specialized equipment (***).

1.4   Measurement

In discussing this parameter, we must first of all define how measurement differs from other evaluation methods. An answer to this question is provided by the following definition, which was given by Professor K.P. Shirokov in GOST 16263–70 [9]:

Measurement is the act of experimentally determining a physical quantity using specialized equipment.

This definition includes four traits:

1. Only physical quantities (i.e., properties of physical objects, phenomena, or processes) can be measured. Thus, sociological, economic, psychological, philological, and other quantitative assessments of non-physical quantities remain outside the bounds of metrology.
2. Measurement is the experimental estimation of parameters, i.e., it always involves an experiment. Thus, determination of a calculation through calculation using a formula and known input data, statistical assessment of product quality indicators via an opinion survey, and other similar procedures cannot be called "measurement."
3. Measurement is performed using special technical equipment – carriers of units of measurement or scales, called measuring instruments. Thus, this definition does not include other determination methods that do not use technical devices (in particular, organoleptic determinations, and determinations via expert assessment).

   It should, of course, be noted that the widespread use of analytical measurements and the increased importance of this measurement field have made it necessary to provide an expanded definition of this trait. The issue is that many analytical measurements are made by performing a series of operations, with the portion of the operation involving use of measurement instruments definitely not determining the accuracy of the result. For example, laboratory measurements of the quality of oil in a railroad tank car will include the following mandatory operations: Collection of a sample, delivery of the sample to the laboratory, preparation of the sample, and measurement.

   The quality with which each of these operations is performed has an effect on measurement accuracy, and an error in performing any of them may have a decisive impact. Strict rules for performing these operations are described in a metrological document called a Measurement Procedure (MP). By analogy with medical terminology, we could say that an MP is a "prescription" describing the measurement procedure, with strict compliance being required. Obviously, in the case of such measurements, it is not so much the measuring instrument itself as the overall MP that plays the decisive role in providing the required measurement accuracy. In such cases, it would be logical to understand the "special technical device" in K.I. Shirokov's definition as referring to the MP as a whole (including the measuring instruments used in the procedure).

4. Measurement is defined as determination of the value of a quantity. Thus, measurement is the "comparison of a quantity against the unit or scale for that quantity." This approach has been developed through hundreds of years of practical experience in measurement, and is completely consistent with the concept of "measurement" defined over 200 years ago by the great mathematician L. Euler: "It is not possible to determine or measure one quantity other than by assuming that another quantity of the same type is known and determining the ratio between the quantity being measured and that quantity" [10].

Measurements may be classified as follows:

1. In terms of the trait of accuracy – "measurements of equal accuracy" and "measurements of unequal accuracy."
   "Measurements of equal accuracy" refers to a series of measurements of some quantity performed using measuring instruments of identical accuracy under identical conditions.
   "Measurements of unequal accuracy" refers to a series of measurements of some quantity performed using measuring instruments with different accuracies and/or under different conditions.
   The methods used for reduction of "measurements of equal accuracy" and "measurements of unequal accuracy" are slightly different. Thus, prior to reducing a series of measurements, a check needs to be performed to determine whether or not the measurements are of equal accuracy. This is performed via a statistical procedure based on the Fisher goodness-of-fit test.

2. In terms of number of measurements – single measurements and multiple measurements.
   A single measurement is a measurement that is only performed once.
   A multiple measurement is a measurement of the same quantity in which the result is obtained from several single measurements (readings) in succession.
   What is the number of measurements above which a measurement can be considered a "multiple measurement"? There is no strict answer to this question. However, it is known that tables of statistical distributions can be used for reduction of a series of measurements based on the rules of mathematical statistics for a number of measurements $n > 4$. It is therefore believed that a measurement can be considered a "multiple measurement" if the number of measurements is at least 4.
   In many cases, especially in everyday life, "single measurements" are performed. For example, a specific instant of time is generally measured using a clock only once.

In the case of some measurements, however, a single reading is not sufficient to provide confidence in the result. In everyday life as well, it is frequently recommended that several measurements be performed rather than a single measurement. For example, when monitoring human arterial blood pressure, the fact that blood pressure is variable means that it is desirable to perform two or three measurements and adopt the median value as the result. Pairs of measurements and triple measurements differ from multiple measurements in that it does not make sense to use statistical methods for assessment of the measurement accuracy.

3. Based on nature of the quantity being measured – static or dynamic.

"Dynamic measurement" is defined as measurement of a quantity whose magnitude varies as a function of time. If there is rapid variation in the magnitude of a measurand, said measurand must be measured in such a way that the time is precisely recorded. Examples include measuring the distance to the ground from an airplane that is descending, or measuring the varying voltage of an electrical current. Essentially, a "dynamical measurement" consists of determining the functional dependence of the measurand on time.

"Static measurement" is defined as measurement of a quantity that is assumed (in the context of a specific measurement task) to remain unchanged for the amount of time required to perform the measurement. For example, measurement of the linear size of a fabricated part at standard temperature can be considered a static measurement, since variations in the shop temperature at a level of a few tenths of a degree will introduce a measurement error of no more than 10 μm per m, which is insignificant compared relative to the error in fabrication of the part. Thus, in the context of this measurement task, the measurand can be considered constant. When performing calibration measurements of the line standard meter on the primary state reference, the temperature is held stable to within 0.005°C; such temperature variations will give rise to a thousand times smaller measurement error – no more than 0.01 \mu m. However, in the context of this measurement task, this would be a substantial variation, and the variation in temperature during the measurement process must be taken into account in order to achieve the desired measurement accuracy. These measurements should therefore be performed using a procedure appropriate to dynamic measurements.

4. Based on purpose of measurement – technical measurements and metrological measurements.

"Technical measurements" are measurements intended to obtain information on the properties of physical objects, processes, and phenomena in the surrounding world. Such measurements are performed, for example, when monitoring and controlling scientific experiments, when monitoring the parameters of products or various technological processes, when controlling the movement of various types of vehicles, for diagnosis of disease, for monitoring environmental pollution, etc. Technical measurements are generally performed using ordinary measuring instruments. However, measurement standards are frequently used in the performance of experiments involving high-precision or unique measurements.

"Metrological measurements" are measurements to ensure metrologic trace-ability and required accuracy of technical measurements. Metrological measurements include the following:

- Reproduction of units and scales for physical quantities using primary measurement standards and dissemination of said units and scales using lower-accuracy measurement standards;
- Calibration of measuring instruments;
- Measurements performed during verification of measurement instruments;
- Other measurements performed for this purpose (for example, measurements made during comparison of measurement standards with the same accuracy level), or for addressing other requirements internal to metrology (for example, measurements for more accurate determination of fundamental physical constants and for improvement of standard reference data for the properties of substances and materials; and measurements for confirmation of the stated capabilities of laboratories).

Metrological measurements are performed using measurement standards. Products intended for consumption (by industry, agriculture, the army, government executive-branch agencies, the public, etc.) are obviously produced using technical measurements, while the system of metrological measurements serves as an infrastructure for the system of technical measurements, and is required in order for the latter system to exist, be developed, and be improved.

5. Based on the dimensions of the units used – absolute and relative.
   A "relative measurement" is a measurement of the ratio of some quantity to a quantity of the same dimensions acting as a unit. For example, determination of the activity of a radionuclide in a source by measuring the ratio of the activity to that of the radionuclide in another source that has been certified as a standard for this quantity.
   The opposite concept is "absolute measurement." When such a measurement is being performed, no unit for the quantity being measured is available to the experimenter. Therefore, it is necessary to reproduce the unit during the measurement process itself. This is possible via two methods:

   - obtaining it "directly from the natural world," i.e., reproducing it based on the use of physical laws and fundamental physical constants (in the VIM International Vocabulary of Metrology [11], this type of measurement is called a "fundamental measurement");
   - reproducing the unit based on a known relationship between the unit and the units for other quantities.

   Thus, we can define an "absolute measurement" as follows:
   Absolute measurements are measurements based on direct measurements of one or more fundamental quantities and/or the use of fundamental physical constants.
   For example, measurement of force using a dynamometer would be "relative measurement," while measurement of force using the physical

constant (the universal gravitational constant) and measures of mass (a fundamental quantity in the SI system) g.

Implementation and metrological traceability of relative measurements is generally the best solution for many measurement tasks, since relative measurements are simpler, more accurate, and more reliable than absolute measurements. In practice, absolute measurements (in the sense that corresponds to the meaning of "fundamental measurement") should be used as an exception. Absolute measurements are used for independent reproduction of SI fundamental units and discovery of new physical relationships.

6. Based on method by which measurement results are obtained – direct measurements, indirect measurements, measurements in a closed series, and combined measurements.

Direct measurements are measurements performed using measuring instruments that store a unit or scale for the quantity being measured. Examples include measurement of the length of a part using a micrometer, measurement of current using an ammeter, or measurement of mass on a scale.

Indirect measurements are measurements in which the value of a quantity is determined based on results for direct quantities functionally related to the quantity being determined. Examples of indirect measurements:

– Determination of the height h of an object based on the results from measurements of the distance l to the object and the angle α of a right triangle including the object, which are related by the equation $h = l \times \tan \alpha$;
– determination of the density ρ of a homogeneous cylindrical body based on the results from measurement of the mass m and the height h and diameter d of the cylinder, which are related by the equation

$$\rho = \frac{m}{0.25\pi d^2 h}.$$

Measurements in a closed series are measurements of several homogeneous quantities performed simultaneously, in which the values of these quantities are determined by solving a system of equations obtained via measurements of various combinations of these quantities.

The classical example of measurements in a closed series is calibration of a set of weights using a single standard weight, by measuring various combinations of weights from the weight set being calibrated, and solving the resulting set of equations.

Combined measurements are simultaneous measurements of two or more non-homogeneous quantities to identify dependences between them. In other words, combined measurements are measurements of the dependences between quantities.

One example of a combined measurement would be measurement of the coefficient of linear thermal expansion (CLTE), which is performed by simultaneously measuring the change in temperature of a sample of the material under test and the corresponding increase in the length of that sample, and then performing mathematical reduction of the measured results.

Distinctions should also be made with respect to measurement fields, types, and subtypes.

Measurement field is understood to mean the full set of measurements of physical quantities inherent in some area of science and engineering and distinguished by specific characteristics.

The following measurement fields are currently identified:

- Measurement of quantities related to space and time;
- Mechanical measurements (including measurements of kinematical and dynamical quantities, measurements of the mechanical properties of substances and materials, and measurements of the mechanical properties and shapes of surfaces);
- Thermal measurements (thermometry, measurements of thermal energy and the thermal-physics properties of substances and materials);
- Electrical and magnetic measurements (measurements of electric and magnetic fields, measurements of the parameters of electric circuits, measurements of the characteristics of electromagnetic waves, and measurements of the electrical and magnetic properties of substances and materials);
- Analytical (physical and chemical) measurements;
- Optical measurements (measurements of quantities in physical optics, coherent optics, and nonlinear optics, measurements of the optical properties of substances and materials);
- Acoustic measurements (measurements of quantities in physical optics, and measurements of the acoustic properties of substances and materials);
- Measurements in atomic and nuclear physics (measurements of ionizing radiation and radioactivity, and measurements of the properties of atoms and molecules).

"Type of measurement" is a subcategory within "field of measurement," with its own unique characteristics, and homogeneous measured quantities.

Within the field of electrical and magnetic measurements, the following [types of measurements] can be identified: Measurement of electrical resistance, measurement of electrical voltage, measurement of electromagnetic field (EMF), measurement of magnetic induction, etc.

"Subtype of measurement" is a subcategory within "type of measurement," which describes the characteristics for measurements of a homogeneous quantity (range, magnitude of quantities, measurement conditions, etc.).

For example, with respect to length measurements, a distinction is made between measurement of large lengths (tens, hundreds, or thousands of kilometers) and measurement of small and extremely small lengths.

1.5 Measurement Method and Procedure

The solution to any measurement task involves implementation of a specific measurement principle.

A measurement principle is the physical phenomenon or effect on which measurements using a specific measuring instrument are based. Examples of measurement principles are as follows:

- Use of the Josephson effect for measurements of electrical voltage;
- Use of the Doppler effect for measurements of [radial] velocity;
- Use of the force of gravity to measure mass by weighing;
- Resistance of platinum as a function of temperature, as implemented in platinum resistance thermometers;
- ThermoEMF as a function of temperature difference, as implemented in thermoelectric thermometers.

However, selection of a measurement principle does not completely define the measurement method, since measurement method is a much more general concept that describes the method used to solve a measurement task. It is defined as follows:

A measurement method is a procedure or set of procedures (consistent with the measurement principle implemented) for comparison of the quantity being measured against the unit or scale for said quantity.

A wide variety of measurement methods are used. They may be classified based on a variety of criteria. The first criterion is the physical principle used. In terms of physical principle, measurement methods are divided into electrical, magnetic, acoustic, optical, mechanical, etc. The second criterion used is the way in which the measurement signal varies as a function of time. According to this criterion, all measurement methods are divided into static methods and dynamic methods. The third criterion is the means by which the measuring instrument interacts with the subject of the measurements. In terms of this criterion, measurement methods are divided into contact methods (where the measuring instrument sensor is in contact with the subject of the measurements) and non-contact methods (where the measuring instrument sensor does not come into contact with the subject of the measurements). The fourth criterion is the type of measurement signal used in the measuring instrument. According to this criterion, measurement methods are divided into analog methods and digital methods.

This classification scheme could undergo further development. However, a metrological classification scheme for measurement methods based on the methods used for comparison of the measured quantity against the unit (see Fig. 2) would be more general. According to this criterion, all measurement methods fall under one of two methods:

- Direct estimation method (the value of the quantity is determined directly from the readout device on the measuring instrument, i.e., a clock or ammeter);
- Method of comparison against a measure (the quantity being measured is compared against a quantity that reproduces a measure, e.g., measurements of mass using a lever scale).

There are several versions of the method involving comparison with a measure: the differential method, the substitution method, the supplementation method, and the coincidence method.

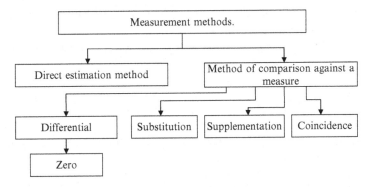

Fig. 2 Metrological classification of measurement methods

The differential method is a measurement method in which the difference between the quantity being measured and a homogeneous quantity with a known value that differs only slightly from the value of the quantity being measured.

An example of a differential method is calibration of length measures by comparison with standard measures using a comparator (an instrument designed for comparison of measures) In this case, the measured quantity X is partially balanced against the quantity reproducing the measure X_m, and the difference ΔX is determined . Thus, the result of the measurement is $X = X_m + \Delta X$. The differential method enables the accuracy of measurement to be substantially increased. For example, if $\Delta X = 0.01\%$, and the relative measurement error ΔX is 1%, the relative error in the measured result X will be 0.01% (if the error in the measure is ignored).

One special case of the differential method is called the zero-measurement method – a measurement method in which the resultant effect of the quantity being measured and of the measure on the comparator is reduced to zero. In this case, the value of the quantity being measured is equal to the value reproduced by the measure. Examples of the zero method include: weighing a mass on a scale using a set of weights; measurement of electrical voltage using a fully balanced bridge.

The differential method reduces the random error of measurement. The substitution method is useful in dealing with systematic errors. *The substitution method is a comparison-with-a-measure method in which the quantity to be measured is substituted with a quantity reproduced by a measure.* Since both measurements are performed by the same instrument under identical conditions, systematic error can largely be compensated for. For example, a significant component of the measurement error when determining mass on lever scales – the error from unequal scale arms – can be eliminated from the measured result if the measurement is performed using Borda's method (weighing by alternately placing weights and the unknown mass on the same scale pan).

In certain measurement tasks, it is convenient to use other versions of the comparison-against-a-measure method: The supplementation method and the coincidence method. *The supplementation method is a comparison-against-a-measure method in which the quantity being measured is supplemented by a measure so that the comparison instrument operates on the sum, which is equal to a previously known value.* For example, in certain cases, it may turn out that a mass measurement in which a weight whose value is known to high accuracy is balanced against the mass to be determined and a set of lighter weights placed in the other pan of the balance.

The coincidence method is a measurement method in which the difference between the quantity being measured and the quantity reproducing a measure is determined via coincidence of scale markings or periodic signals. One example of this method is measurement of length using a vernier caliper. The coincidence method is frequently used in measurements of periodic processes. If, for example, a frequency v is to be measured using an oscillator with a standard frequency v_0, the coincidence method would consist of recording the number n of the frequency v coincident in time with the number n_0 of frequency v_0. In this case, the result of the frequency measurement, which is given by the equation $v = v_0 n/n_0$, will have almost the same error as the error in reproduction of the standard frequency v_0 by the oscillator.

Obviously, selection of a measurement method will depend on theoretical justification, availability of the necessary measuring instruments, and the type and design characteristics of the measuring instruments (measure, measuring device, etc.). For example, an extremely simple measurement task such as measuring the height of a factory smokestack may be performed by selecting one of the following methods:

- climbing the stack with a ruler, and then performing the measurement (method of comparison against a measure);
- Flying a helicopter with an altimeter until it is level with the smokestack, and measuring the altitude (direct-assessment method);
- Calculating the height of the stack as the leg of a right triangle based on measurements of the distance to the stack and the angle of this triangle (indirect measurement).

If the measurement method calls for development of basic procedures for use of the measuring instruments, the measurement procedure is essentially the measurement process that will best implement the selected measurement method. *"Measurement procedure" (MP) is the term used to describe the required set of rules and operations such that compliance with said rules and operations will ensure that the required measurement results are obtained using the selected method.* The MP includes requirements with respect to selection of measuring instruments, specifications regarding preparation of the measuring instruments for use, requirements with respect to measurement conditions, specifications regarding measurement procedures and reduction of measurement results, including estimation of measurement accuracy. The MP for analytical measurements also includes requirements with respect to sample collection, sample storage, transportation of samples to the measurement laboratory, and preparation of samples for measurement.

Standardization of MPs is of great value for metrological traceability. For this reason, MPs for recurrent measurements are generally covered by a regulatory document of some kind.

1.6 Measuring Instruments

Technical devices that are used in measurements and which have standardized metrological properties are called measuring instruments. The heart of this definition, which describes the metrological essence of a measuring instrument (MI), is contained in the words "standardized metrological properties." The existence of "standardized metrological properties" means, first of all, that the MI is capable of storing or reproducing a unit (or scale), and second, that the magnitude of this unit remains unchanged for a specified time. If the magnitude of the unit were not stable, it would be impossible to guarantee the required accuracy of the measurement result. This then implies three conclusions:

- It is possible to make a measurement only when the device intended for this purpose is able to store a unit such that the magnitude of the unit is sufficiently stable (invariant as a function of time);
- Immediately after fabrication, the technical device is still not an MP; it becomes one only after a unit is transferred to said technical device from another, more accurate MI (this operation is called calibration);
- The magnitude of the unit stored in the MP must be periodically verified, and its prior value restored, if necessary, via a new calibration.

In terms of purpose, a distinction is made between ordinary MPs used for performing technical measurements and metrological MPs intended for performance of metrological measurements. Metrological MIs are called measurement standards.

In order to understand the metrological classification of MPs, consider the typical block diagram for a direct measurement shown in Fig. 3 [7].

A direct measurement procedure consists of the following elementary operations:

- Transformation $Q = F(X)$ of a measured quantity X into another quantity Q that is either homogeneous or inhomogeneous with respect to the first quantity;

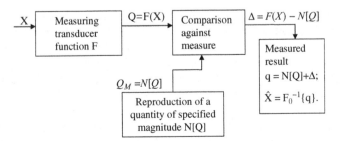

Fig. 3 Structural diagram of direct measurement

- Reproduction of a quantity $Q_m = N[Q]$ of specified magnitude, approximately equal to the magnitude of the quantity Q (where $[Q]$ is the unit for said quantity);
- Comparison of the homogeneous quantities Q and Q_m, determination of the difference between them $\Delta = Q - Q_m$, and determination of the actual transformation function for the quantity being measured $F[X] = N[Q] + \Delta$;
- Generation of the measurement result $\hat{X}$ by comparison against the calibration function for the MP $Q_0 = F_0(X)$, which acts as a memory element. At this point, the back-transformation $\hat{X} = F_0^{-1}\{N[Q] + \Delta\}$ is performed using the transformation function $F_0(X)$ that the MI had at the time the calibration relationship was determined. Obviously, the magnitude $[Q_0]$ of the unit determined in the last calibration may differ from the magnitude $[Q]$ of the unit stored in the MI at the time of measurement. This is one of the major sources of measurement error.

All of the above operations are performed using technical devices that are either standalone measurement instruments or are part of an MI.

1.6.1   Measuring Transducers

This first operation requires a *measuring transducer* (MT) – *an MI intended for transformation of a measured quantity into another quantity or into a measurement signal convenient for processing, further transformation, display, or transmission.* A distinction is made between primary and intermediate MTs based on location within the measurement chain. *A primary MT, also called a sensor, is an MT on which the quantity being measured acts directly. Other MTs are called intermediate. They are located downstream of the primary MT, and may perform a variety of operations related to transformation of the measurement signal. These operations generally include the following:*

- A change in the physical type of the quantity;
- A scale transformation (linear or non-linear);
- A scale and time transformation;
- An analog-digital transformation;
- A digital-analog transformation;
- A functional transformation (any mathematical operations on quantity values).

It should be remembered that this classification is somewhat arbitrary. First of all a single MI may include several primary MTs (e.g., a thermocouple as part of a thermoelectric thermometer). Second, the specialized nature of analytical measurements also causes this classification principle to break down.

Analytical measurements are a transformation of the measured quantity and comparison of this quantity against a measure, with being an information-bearing parameter describing the medium being analyzed (an information-bearing parameter is a parameter that contains information on the measured quantity). Such measurements are generally performed using a set of MTs of the following types [12]:

– MT$_1$: Composition–composition MTs that enable large-scale transformation of probes for analysis. The sample is characterized by an information-bearing parameter C (concentration of the component being measured) and a combination of non-information-bearing parameters $\bar{C}_n$, including the contribution of undetermined (interfering) components and the thermodynamic parameters of the medium being analyzed. Passage through an MT1 includes processes such as cleaning, drying, and modifying the temperature and pressure of the mixture to the required values, and, after these transformations have been performed on the medium being analyzed, collection of the required amount of medium. MT1s are generally called sampling units or sample preparation units.

– MT$_2$: A composition–property MT that supports transformation of the quantity being measured, C, into some physical or chemical property suitable for later measurement and recording. In many cases, this transformation occurs in two phases: Production of a liquid-phase or solid-phase intermediate product in which the concentration of the component is $Y_{int}(C)$, followed by transformation of this concentration into the property $\Phi\,(Y_{int})$.

– MT$_3$: An MT of the property–input signal type, which supports transformation of the measured quantity into a measurement-signal output W. This transformation is also generally performed in two stages: Into an intermediate signal $W_{int}(\Phi)$ and then into the output signal $W(W_{int})$. In this case, the transformation of W_{int} into W involves transformation of one electrical signal into another.

After obtaining output signals from the item being analyzed using a set of MTs, a calibration function is used to compare the measured quantity against the measure and generate estimation values C^* of the quantity C being measured.

This set of MTs does not fit within the classification presented above, since the measured quantity affects not only the first MT in the measurement chain, but also affects all of them, including MT1, MT2, and the first transducer in the MT3 group. In this case, only the second transducer in the MT3group is an intermediate transducer. This then implies that in analytical instruments, the entire set of MTs plays the role of a primary MT, successively transforming, in several phases, the measured quantity into a measurement signal.

1.6.2 Measures

Reproduction of a quantity $Q_m = N[Q]$ of specified magnitude is accomplished using a measure of the quantity Q. *An MT intended for reproducing and/or storing the physical magnitude of one or more pre-specified magnitudes whose values are known to the requisite accuracy.* Examples of such measures include the line standard meter, a normal element (a measure of EMF with a nominal value of 1 V), a quartz oscillator (a measure of the frequency of electrical oscillations), a source of microscopic flows of gasses and vapors (an ampule containing material released in gaseous form that is a measure for rate of transformation into the target substance gas. Measures are divided into *single-valued measures* (*measures storing a single magnitude of a quantity*, for example, a plane-parallel end measure of length or a constant-capacitance capacitor) and *multi-valued measures* (*a measure storing several magnitudes of a quantity*, e.g., a line standard meter and a variable-capacitance capacitor). In terms of practical measurements, wide use is made of *sets of measures* (*sets of measures containing various magnitudes of the same quantity*, e.g., a set of plane-parallel end measures of length) and *boxes of measures* (*sets of measures structurally combined into a single device with devices for connecting the measures in various combinations*, e.g., an electrical resistance box) in addition to single measures.

By reproducing or storing the magnitude of a quantity assigned a certain value, a measure stores the unit for that quantity. In other words, measures carry the units for quantities, and is therefore serves as the basis for measurement.

Standard samples are a special class of measure. *A standard sample is a measure of one or more quantities characterizing the composition or properties of a substance or material, in the form of a sample of said substance or material.* Examples of standard samples for properties include standard samples of benzoic acid as measures for heat of combustion, standard samples of special steel as measures for the properties of ferromagnetic materials, and standard samples of quartz as measures for relative dielectric permittivity. Various substances, e.g., metals and alloys with well-defined values of the dominant component and impurities present, are examples of standard samples for composition. The primary application for standard samples is calibration of MIs when performing analytical measurements. In addition, like ordinary standards, homogeneous standard samples are used to create hierarchical (in order of decreasing accuracy) chains for dissemination of units. Standard samples are used in these chains for dissemination of units to less accurate MIs, including other standard samples.

Special standard samples called calibration gas mixtures (CGMs), which are very widely used in gas analysis measurements, have a special status in the system of measures. *A calibration gas mixture is a cylinder containing a pure gas or gaseous mixture, and certified by a metrological service as a unique measure for the concentration of components in a gaseous mixture.* There are certain differences between CGMs and standard samples in the form of solid objects. The most important of these is that they are consumed in the process of measurement, and this in turn frequently leads to a substantial increase in the cost of multiple measurements.

As a general rule, standard samples are prepared and distributed by measurement laboratories based on orders received from specialized forms or metrological laboratories. However, there are exceptions. For example, some samples can be prepared by the person performing the measurement, provided preparation is in strict compliance with the preparation requirements set forth in the specifications for the standard samples. For example, a standard sample for the concentration of iron in water may be prepared by dissolving a standard sample of powdered iron into a specified quantity of distilled water. Such standard samples are frequently called calibration mixtures.

In conclusion to our discussion of measures and their use, we should also mention that it is also possible to use certain natural phenomena as measures. For example, precision measurements of angles widely rely on a natural standard – the full angle (planar or solid). The light from one specific star in the sky has been used for metrologic traceability of photometers. The relative spectral energy distribution in the spectrum of this star observed from various points in the USSR was carefully determined. Once the energy curve of this star was found to be constant, this methodology was legally established for calibration of ultraviolet photometers.

The last example illustrates how instruments can be calibrated using tables of standard reference data. These calibrations include the following:

- In mechanical measurements – the mechanical characteristics of various substances (for example, the density of pure substances at specified temperature, humidity, and pressure);
- In temperature measurements – the constants characterizing phase transitions (melting/hardening or boiling/condensation), EMFs of various thermocouples, etc.;
- In electrical measurements – the characteristics of various stable electrical phenomena (for example, the EMFs of various galvanic pairs);
- In optical measurements – various atomic constants, since all physical optics is based on the emissive and absorptive properties of atoms and molecules;

Analytical measurements very commonly make use of standard reference data. Such data includes information on a wide variety of properties of pure substances, various dependences of the properties of alloys and gaseous mixtures on composition, absorption coefficients and indices of refraction of transparent substances, hygrometric and psychrometric tables, etc.

The standard reference data category is one of the most important in metrology. Scientific research in this area is being performed by all of the large metrological institutes in the world and many physical laboratories in various countries. The US metrological center – the National Institute of Standards and Technology (NIST), which has as one of its primary missions development and approval of standard reference data – is playing a leading role. In Russia, coordination of research in this area is performed by the Scientific Research Institute for Standard Reference Data (NIISSD) under the Russian Federation Federal Agency for Technical Regulation and Metrology. On the world level, research is coordinated by the Committee on Data for Science and Technology (CODATA).

1.6.3 Measuring Devices

A measuring instrument that performs all of the actions indicated in Fig. 3 is called a measuring device. *A measuring device is a measuring instrument intended for measurement of a measurand over an established range.* As a general rule, a measuring device includes devices for transformation of a measurand into a measurement data signal and a device for displaying the data in the most accessible form.

In many cases, the display device has a scale and an arrow, a diagram, a digital annunciator panel, or display, which can be used for reading or recording the result of the measurement. In computerized measuring instruments, the measurement result can be automatically recorded on various specific media. The following types of measuring devices are distinguished: analog devices (the output signal is a continuous function of the measurand) and digital devices (the output signal is represented in digital form), display devices (which only allow the readings to be read) and recording devices (which allow the measured results to be recorded), summing devices (the readings are functionally related to the sum of two or more quantities) and integrating devices (the value of the measurand is determined by integration over another quantity). For example, micrometers and digital voltmeters are classified as display measuring devices, while barographs are classified as recording measuring devices.

A distinction is also drawn between direct-action devices and comparison devices. In a direct-action device, the measured result is obtained directly from the device display. Examples of such devices include the ammeter, manometer, and mercury glass thermometer. Direct-action measuring devices are intended for direct-estimation measurements.

In contrast with these devices, comparison-with-a-measure measurements are performed using comparison measuring devices, which are also called comparators. *A comparison measuring device is a measuring device intended for direct comparison of a measurand to a quantity whose value is known.* Examples of comparators include: Dual-pan balance, interference-based length comparator, electrical resistance bridge, electrical measuring potentiometer, and a photometric bench with photometer. Comparators need not store a unit in order to perform their functions. Strictly speaking, such comparators cannot be considered measuring instruments. They must nevertheless have several important metrological properties, with the emphasis being placed on low random error and high measurement sensitivity.

Any direct-acting measuring device may be used as a comparator if it is used for successive measurement of a quantity reproducing a measure and an unknown quantity. This is precisely what is done in analytical measurements: Standard samples are generally certified by comparison, using a comparator consisting of any analytical device capable of measuring to acceptable accuracy the quantity reproduced by these samples. As a general rule, using a measuring device in this way can provide a much higher measurement accuracy than when it is used for direct-estimation measurements. This is easy to understand in light of the fact that using a measuring device as a comparator implements the measurement method called the "substitution method," which eliminates the systematic error introduced by the measuring device. On the other hand, in the substitution method, the measured result is burdened with twice the random error, since two measurements are performed (of the measurand and the quantity reproducing the measure). Thus, if a measuring device has a high random error, it is not suitable for use as a comparator. If, on the other hand, the random error is small, the measurement error will be smaller than for measurement by direct estimation (due to the absence of the systematic component).

1.6.4 Measuring Installations and Measuring Systems

Sets of measuring instruments are frequently combined into units called measuring installations or measuring systems. *A measuring installation is a set of functionally combined measures, measuring devices, measuring transducers, and other devices intended for measurement of one or more quantities and located in a single location.* A measuring installation and the standards included in it is called a calibration or standards installation, while an installation used for testing a product of any type is called a test bench. Some types of large measuring installations are called measuring machines.

A measuring system is a set of functionally connected measures, measuring devices, measuring transducers, computers, and other equipment located at various points throughout the area being monitored and used for measurements of one or more quantities related to said area. For example, the measuring system for a heat and electricity generating plant enables measured data to be obtained with respect to several quantities in various power generating units; changes in weather are monitored using a measuring system consisting of several functionally interrelated measuring units that are spatially separated by significant distances. Measuring systems are classified based on purpose into measuring and information systems and measuring and control systems, etc. A distinction is made between one-channel, two-channel, three-channel, etc., measurement systems, depending on the number of measurement channels.

Chapter 2
Measurement Errors

2.1 Classification of Measurement Errors

The effectiveness of the use of measurement information depends on the *precision of the measurements – the properties that reflect the closeness of measurement results to the true values of the measured quantities.* Measurement precision can be greater or lesser, depending on allocated resources (expenditures for measuring instruments, conducting measurements, stabilizing of external conditions, and so forth). It is obvious that this should be optimal: sufficient to complete the appointed task but no more, since further increase in precision leads to unjustified financial expenditures. Hence along with the concept of precision is also used the concept of *the certainty of measurement results, by which is understood that the measurement results have a precision that is sufficient to solve the task at hand.*

The classical approach to evaluating accuracy of measurement, first applied by the great mathematician Karl Gauss and then developed by many generations of mathematicians and metrologists, can be presented in the form of the following sequence of affirmations.

1. The purpose of measuring is to find *the true value of a quantity – the value that ideally would characterize the measurand, both qualitatively and quantitatively.* However, it is in principle impossible to find the true value of a quantity. But not because it does not exist; any physical quantity inherent in a concrete object of the material world has a fully defined magnitude, the ratio of which to the unit value is the true value of this quantity. This signifies no more than the unknown of the true value of a quantity, which is in the gnoseological sense an analog to absolute truth. The best example to confirm this position is the set of fundamental physical constants (FPCs). They are measured by the most authoritative scientific laboratories of the world, with the highest accuracy, and then the results obtained by different laboratories are coordinated with each other. In this, the coordinated FPC values are established with such a large number of significant digits that any change in successive refinement would occur only in the last significant digit. Hence, the true values of the FPCs are unknown, but

A.E. Fridman, *The Quality of Measurements: A Metrological Reference*,
DOI 10.1007/978-1-4614-1478-0_2, © Springer Science+Business Media, LLC 2012

each succeeding refinement makes the value of this constant as derived by the world community approach its true value.

In practice, rather than the true value, there is used *the conventional true value – the value of the quantity that is derived experimentally and is so close to the true value that it can be used instead of it in the measurement task set forth.*

2. *Deviation of the result X from the true value X_{tr} (the conventional true value X_{ctr} of a quantity is called the measurement error*

$$\Delta X = X - X_{tr}(X_{ctr}). \qquad (2.1)$$

Due to the imperfection of the methods used and measuring instruments, the instability of measurement conditions, and other reasons, the result of each measurement is burdened with error. But, since X_{tr} and X_{ctr} are unknown, the error ΔX likewise remains unknown. It is a random variable, and thus in the best case can only be estimated according to the rules of mathematical statistics. This absolutely must be done, since the measurement result has no practical value without indicating an error estimate.

3. Using different estimation procedures, an interval estimate of the error ΔX is found, in the form that most often provides *confidence intervals* $-\Delta_P, +\Delta_P$ of *the measurement error for a specified probability P. These are understood to be the upper and lower bounds of the interval within which the measurement error ΔX is located with a specified probability P.*

4. It follows from the preceding fact that

$$X - \Delta_P \leq X_{tr}(X_{ctr}) \leq X + \Delta_P \qquad (2.2)$$

– the true value of the measurand is located, with probability P in the interval $[X - \Delta_P; X + \Delta_P]$. The bounds of this interval are called the confidence limits of the measurement result.

Hence, a measurement result finds not the true (or conventional true) value of the measurand, but an estimate of its value in the form of the limits of an interval where it is located with the specified probability.

Measurement errors can be classified by various criteria.

1. They are divided into absolute and relative errors according to their method of expression. *An absolute measurement error is an error expressed in units of the measurand.* Thus, the error ΔX in formula (2.5) is an absolute error. A deficiency of this method of expressing these values is the fact that they cannot be used for a comparative estimation of the accuracy of different measurement technologies. In fact, $\Delta X = 0.05$ mm for $X = 100$ mm corresponds to a rather high accuracy of measurement, while for $X = 1$ mm it would be low. This deficiency is ameliorated by the concept of "*relative error*," defined by the expression (2.7).

$$\delta X = \frac{\Delta X}{X_{tr}} \left(\frac{\Delta X}{X} \right). \qquad (2.3)$$

Hence, *relative measurement error is the ratio of the absolute measurement error to the true value of the measurand or the measurement result.*

To characterize the accuracy, the measuring instrument often uses the concept of *"fiducial error"*, defined by formula (2.8)

$$\gamma X = \frac{\Delta X}{X_\mathrm{n}}, \tag{2.4}$$

where X_n is the value of the measurand, conventionally taken as the normalized value of the scale range of the measuring instrument. Most commonly, the difference between the upper and lower limits of this scale range is used for the X_n.

Hence, *a fiducial error of the measuring instrument is the ratio of the absolute error of the measuring instrument at a given point in the scale range of the measuring instrument to the normalized value of this range.*

2. Measurement errors are divided into instrumental, methodological, and subjective, according to the source of the measurement errors.

An instrumental measurement error is that component of measurement error that is caused by imperfection in the measuring instrument being used: the divergence of the actual functioning of the instrument's transformation from its calibrated relationship, unavoidable noise in the measurement chain, delay in the measured signal as it passes through the measuring instrument, internal resistance, and so forth. Instrumental error of measurements is divided into *intrinsic error (measurement error when using a measuring instrument under normal conditions) and complementary (the component of measurement error that arises as a consequence of the deviation of any of the influencing variables from its nominal value or exceeding the limits of its normal range of values).* The method of estimating is shown below.

Methodological measurement error is that component of measurement error caused by imperfection in the method of measurement. This includes errors caused by the deviation of the accepted model of the object of measurement from the actual object, imperfection in the method of realization of the principle of measurement, inaccuracy in the formulas used to find the results of measurements, and other factors not associated with the properties of the measuring instrument. Examples of methodological measurement errors are:

- Errors in the manufacture of a cylindrical body (deviation from an ideal circle) when measuring its diameter;
- Imperfection in determining the diameter of a spherical body as the average of the values for its diameter in two perpendicular planes chosen previously;
- Error in measurements as a consequence of a piecewise-linear approximation of the calibration curve of the measuring instrument, when calculating the measurement results;
- Error in the static indirect method of measurements of the mass of petroleum product in a reservoir due to nonuniform density of the petroleum product with respect to the height of the reservoir.

Subjective (personal) measurement error is that component of measurement error caused by the individual features of the operator; i.e. error in the operator's reading of indicators from the measuring instrument's scales. These are evoked by the operator's condition, imperfection of sensory organs, and the ergonomic properties of the measuring instrument. The characteristics of subjective measurement error are determined by taking into account the capabilities of the "average operator" with interpolation within the limits of the scale interval of the measuring instrument. The most well-known and simple estimation of this error is its largest possible value in the form of half the scale interval.

3. Systematic, random, and gross errors are delineated according to the nature of the event.

Gross measurement error (failure) refers to measurement error significantly exceeding the error expected under the given conditions. They arise, as a rule, from mistakes or incorrect actions of the operator (incorrect reading, mistakes in writing or calculations, improper switch-on of the measuring instrument, and so forth). A possible reason for failure may be malfunctions in the operation of the equipment, as well as transient sharp changes in the conditions of measurement. Naturally, gross errors must be detected and removed from the series of measurements. A statistical procedure designed for this will be examined in Sect. 4.1.

The division into systematic and random errors is more substantive.

*Systematic measurement error is that component of measurement error that in replicate **measurements** remains constant or varies in a predictable manner.* Systematic errors are subject to exclusion, as far as possible, using one or another method. The most well-known of these is the correction action on known systematic errors. However, it is virtually impossible to fully exclude a systematic error, and some traces remain *even in corrected measurement results.* These traces are referred to as residual bias (RB). *Residual bias is the measurement error caused by errors in computation and in corrective action or by systematic error to which a correction has not been introduced.*

For example, to exclude systematic measurement error caused by instability of the transform function for an analytical instrument, calibration is periodically performed using measurement standards (verifying gas mixtures or standard samples). However, despite this, at the moment of measurement, there will nevertheless be a certain deviation of the actual transform function of the instrument from the calibrated curve, caused by calibration error and drift of the transform function of the instrument since the time of calibration. The measurement error caused by this deviation is residual bias.

Random measurement error is that component of measurement error that varies randomly (in sign and in magnitude) for repeated measurements of one and the same quantity. There are multiple reasons for random errors: noise in the measuring instrument, variations in its indications, random fluctuations of the parameters of the instrument power supply and the measurement conditions, rounding errors on the readings, and many others. No uniformity is observed in the manifestation of such errors, and they appear with repeated measurements of one and the same quantity as a scattering of measurement results. Hence, an estimation of random measurement errors if possible only if based on mathematical statistics (this mathematical discipline was engendered as the study of methods of processing series of measurements burdened with random errors).

In contrast with systematic errors, it is not possible to exclude random errors from measurement results by corrective action, although it is possible to substantially reduce their effect by conducting multiple measurements.

2.2 Laws of Random Measurement Error Distribution

2.2.1 Some Information from the Theory of Probability

It is known from probability theory that the most complete description of a random variable is provided by its distribution law, which can be presented as two mutually linked forms, called the cumulative (integral) and the differential distribution function. *The* cumulative *integral distribution function $F_\xi(x)$ of a random variable ξ is the name given to a function x, equal to the probability that ξ has a value less then x*:

$$F_\xi(x) = P\{\xi < x\} \tag{2.5}$$

Figure 4 shows the chart of a cumulative integral function.
These obvious properties of $F_\xi(x)$ follow from (2.5):

- $F_\xi(x) \geq 0$ (non-negative function);
- if $x_2 > x_1$, then $F_\xi(x_2) \geq F_\xi(x_1)$ (nondecreasing function of x);
- $F_\xi(-\infty) = 0$, $F_\xi(\infty) = 1$;
- $P\{x_1 < \xi < x_2\} = F_\xi(x_2) - F_\xi(x_1)$. $\tag{2.6}$

From (2.6), one may derive the definition of a *differentiable distribution function*:

$$f_\xi(x) = \lim_{\Delta x = 0} \frac{F_\xi(x + \Delta x) - F_\xi(x)}{\Delta x} = \frac{dF_\xi(x)}{dx}. \tag{2.7}$$

Figure 5 shows the graph of a differentiable distribution function $f_\xi(x)$.
It is evident that $f_\xi(x) \geq 0$ for any x. It follows from (2.7) that $F_\xi(x) = \int_{-\infty}^{x} f_\xi(z)\,dz$ and, consequently, $\int_{-\infty}^{\infty} f_\xi(z)\,dz = 1$.

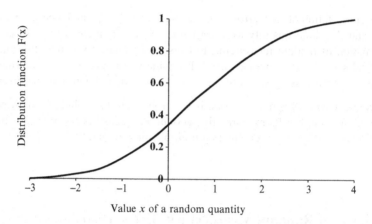

Fig. 4 Cumulative integral distribution function $F_\xi(x)$

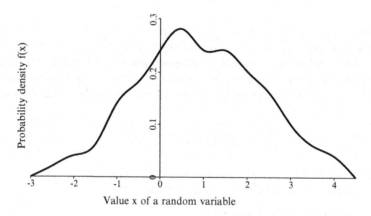

Fig. 5 Differentiable distribution function $f_\xi(x)$

The probability of finding a random variable ξ in the interval $[x_1, \ x_2]$ is equal to

$$P(x_1 \leq \xi < x_2) = \int_{x_1}^{x_2} f_\xi(z) \ \mathrm{d}z. \tag{2.8}$$

It follows from the last equation that the probability that a random variable falls in the specified interval $[x_1, \ x_2]$ is equal to the area under the curve $f_\xi(x)$ between the abscissas x_1 and x_2. If one presents this region in the form of a planar geometric figure, then it will become clear why a differentiable distribution function is called the *probability density of a random variable*.

Every distribution law can be fully characterized by an infinite set of numerical characteristics called the probability moments. The rth order initial moment is calculated from the origin of the coordinates and is defined by the formula

$$\alpha_r = \int_{-\infty}^{\infty} x^r f_\xi(x)\, dx, \quad r = 1, 2, \ldots. \tag{2.9}$$

The most widespread among these is the *first initial moment, referred to as the mean of the random variable*:

$$m_\xi = \alpha_1 = \int_{-\infty}^{\infty} x f_\xi(x)\, dx. \tag{2.10}$$

The mean is the most likely value of the random variable. If, as before, one presents the graph of the probability density as a planar geometric figure, then a point on the axis of the abscissa with coordinate m_ξ will be the center of gravity of this figure. Hence the mean is treated as the center of gravity of the probability distribution.

The central moment is calculated from the mean and is defined by the formula

$$\mu_r = \int_{-\infty}^{\infty} (x - m_\xi)^r f_\xi(x)\, dx. \tag{2.11}$$

The most well-known central moment is the second one, referred to as the *dispersion of the random variable*:

$$D_\xi = \mu_2 = \int_{-\infty}^{\infty} (x - m_\xi)^2 f_\xi(x)\, dx. \tag{2.12}$$

The dispersion characterizes the scattering of the random variable relative to the mean.

A comparatively accurate experimental determination of the third moment requires at least 80 independent measurements, and for the fourth, at least 200. Further increase in the order of moments of a distribution is accompanied by a similarly increasing volume of required measurement information. Hence in practice mainly the first two orders mentioned above are used.

The moments of the distribution are closely associated with the numerical characteristics of the probability distributions. As a rule, two types of numerical characteristics are used: the characteristics of the center of the distribution and the characteristics of the scattering of the random variable.

The center of the distribution can be determined by several methods. The most fundamental method involves determining the mean m_ξ. Another method involves finding the center of symmetry of the distribution, i.e., that point Me on the axis of the abscissa at which the probability of the random variable falling to the left or to the right are the same and equal to 0.5:

$F_\xi(Me) = 1 - F_\xi(Me) = 0.5$. The value Me is referred to as the median or the 50% quantile.

The mode Mo can be used as a center of the distribution; this is the point of the abscissa that corresponds to the maximum of the probability density of the random value $(f_\xi(\text{Mo}) = \max_{x \in (\text{Mo}-z,\, \text{Mo}+z)} f_\xi(x))$. A distribution with one maximum is called unimodal; with two, bimodal; and so forth.

The dispersion D_ξ introduced above is a characteristic of the scattering of a random variable. The dispersion is not always convenient to use, since its dimensionality is equal to the square of the dimensionality of the random variable. Hence, the mean square deviation (MS), equal to the square root of the dispersion taken with a positive sign, is often used in its place:

$$\sigma_\xi = \sqrt{D_\xi}. \qquad (2.13)$$

The mean square deviation is often called the *standard deviation (SD)*.

2.2.2 The Normal Distribution Law

The goal of any measurement is to find the true (actual) value of a measurand. However, the experimenter does not have at hand the set of all possible values of the random variable (called the general set), but a sampling from this set, which incorporates a limited number of measurement results. The numerical characteristics of this sample provide a representation of the characteristics of the center of distribution of the general set – the mean. However, due to the random nature of the sample, they themselves are random variables, and using them for the mean introduces additional error. Hence, it is essential to select from among them the best and most efficient estimate.

In principle, the center of distribution of a sample $(x_1, x_2, ..., x_n)$ of n measurement results can be characterized using the following methods:

– arithmetic mean

$$\bar{x} = \frac{1}{n} \sum_{i=1}^{n} x_i$$

(the mean, taken from the sample);
– median

$$Me = \begin{cases} x_{(n+1)/2}, \ldots n \text{ odd,} \\ \dfrac{x_{n/2} + x_{(n+1)/2}}{2}, \ldots n \text{ even;} \end{cases}$$

- mode *Mo* (value at which the density is maximal);
- geometric mean $g = \sqrt[n]{x_1...x_n}$;
- power mean

$$u = \left(\frac{1}{n} \sum_{i=1}^{n} x_i^F \right)^{1/F},$$

including its particular cases:
- if $F = 1$, the mean,
- if $F = 2$, the mean square

$$S = \sqrt{\frac{1}{n} \sum_{i=1}^{n} x_i^2},$$

- if $F = -1$, the harmonic mean

$$h = n \left(\sum_{i=1}^{n} x_i^{-1} \right)^{-1}.$$

Which of these estimates is the best approximation of the mean of the distribution of a random error? The answer to this question was provided 200 years ago by Carl Gauss. In his work "On the Motion of Celestial Bodies", published in 1809, he formulated three postulates [13].

1. *In a series of independent observations, errors of different sign appear equally frequently.*
2. *Large deviations from the true value occur less frequently than small ones.*
3. *If any value is determined from many observations that are produced under equal conditions and with the same attentiveness, then the arithmetic mean from all observed values will be a more probable value.*

From these axioms, Gauss drew out the distribution law of random values, which has had immense application in science and technology. Let us introduce this proof. Let x_1, x_2, ..., x_n be the results of n uniformly accurate measurements of a quantity, the true value of which is equal to z. Random measurement errors $\varepsilon_i = x_i - z$ are distributed according to some unknown law with probability density $f(x_i - z)$. In accordance with the second postulate, the probability of larger values for the random error are small in comparison with the probability of small values. Hence the probability that the measurement result is equal to x_i is equal to $P_i = \int_{x_i}^{x_i + \Delta x} f(y - z) \, dy \cong f(x_i - z) \, \Delta x$, where Δx is a small value. The probability of obtaining a series x_1, x_2, ..., x_n of measurement results will, by the theorem for multiplying probabilities, be equal to the product of the P_i probabilities:

$$P = P_1 \times P_2 \times ... \times P_n \cong f(x_1 - z) \times f(x_2 - z) \times ... \times f(x_n - z) \times (\Delta x)^n.$$

There exists a $z = \theta$ for which this probability has the maximum value $P_{\max}$. Since $\ln P$ is a monotonic function of P, for $z = \theta$ the function $\ln P$ likewise has a maximum value:

$$\max[\ln P] = \ln P_{\max} = \left\{ \sum_{i=1}^{n} \ln[f(x_i - z)] \right\}_{z=\theta} + n \ln(\Delta x).$$

From this it follows that

$$\frac{d \ln P}{dz}\bigg|_{z=\theta} = \sum_{i=1}^{n} \frac{f'(x_i - \theta)}{f(x_i - \theta)} = 0, \qquad (2.14)$$

where $f'(x_i - \theta)$ is the derivative of the density $f(x_i - \theta)$ on x_i.

In accordance with the third postulate, $\theta = 1/n \sum_{i=1}^{n} x_i$. This expression can be written as $\sum_{i=1}^{n} (x_i - \theta) = 0$. Joining this with (2.14), we derive:

$$\sum_{i=1}^{n} \left[\frac{f'(x_i - \theta)}{f(x_i - \theta)} + C(x_i - \theta) \right] = 0.$$

For this expression to be valid for any values of x_i, it is necessary and sufficient that for any $i = 1, 2, \ldots, n$ this condition be fulfilled:

$$\frac{f'(x_i - \theta)}{f(x_i - \theta)} = -C(x_i - \theta). \qquad (2.15)$$

Expression (2.15) is a differential equation with separable variables relative to the unknown function $f(x - \theta)$. Its solution is the function

$$f(x - \theta) = e^{-C(x-\theta)^2/2} \sqrt{\frac{C}{2\pi}}.$$

Further, substituting this expression into (2.10) and (2.12), after transformations we derive the fact that the mean m of this distribution is equal to θ and the dispersion $\sigma^2 = 1/C$. Taking this into account, the probability density x of the measurement results is equal to

$$f(x) = \frac{1}{\sqrt{2\pi}\sigma} e^{-(x-m)^2/2\sigma^2}. \qquad (2.16)$$

Since the measurement error $\varepsilon = x - m$, its distribution has the form

$$f(\varepsilon) = \frac{1}{\sqrt{2\pi}\sigma} e^{-\varepsilon^2/2\sigma^2}. \qquad (2.17)$$

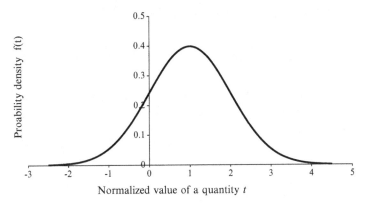

Fig. 6 Graph of the normal distribution density for $m = \sigma$ and $t = \frac{x}{\sigma}$

This function, which describes the distribution of random measurement errors, differs from (2.16) in that it has a mean of zero.

Distribution (2.16) has received the designation of the normal distribution of a random value. It is also referred to as the Gaussian distribution.

In accordance with the central limit theorem of probability theory, the sum of n independent random variables, each of which is small compared with the sum of the other variables, approaches the normal distribution as $n \to \infty$. This provides the foundation for thinking that the normal law is not an artificial mathematical construct, but a fundamental governing law of the phenomena of nature and the material world.

Figure 6 shows a chart of the normal probability density. It has the following properties.

1. The distribution is symmetric with respect to the mean: $f(x - m) = f(-x + m)$. Hence $Me = Mo = m$; the median and mode of the distribution coincide with the mean.

2. The distribution is reproducing. This means that the sum of values distributed normally is likewise distributed normally. This property of a normal distribution facilitates the creation of a family of probability distributions that are widely used in the statistical processing of measurement results.

3. $P\{a \leq x < b\} = \int\limits_{a}^{b} f(x)\,dx = F\left(\frac{b-m}{\sigma}\right) - F\left(\frac{a-m}{\sigma}\right),$ \hfill (2.18)

where $F(x) = \frac{1}{\sqrt{2\pi}} \int\limits_{-\infty}^{x} e^{-0,5t^2}\,dt$, is the cumulative integral function of a normal distribution.

The definite integral $\Phi(x) = \frac{1}{\sqrt{2\pi}} \int\limits_{0}^{x} e^{-0.5t^2}\,dt$ is called the Laplace integral function. The following equalities are valid for this: $\Phi(x) = F(x) - 0.5$; $\Phi(x) = -\Phi(-x)$; $\Phi(-\infty) = -0.5$; $\Phi(0) = 0$; $\Phi(\infty) = 0.5$. Hence the other notation (2.18): $P\{a \leq x < b\} = \Phi\left(\frac{b-m}{\sigma}\right) - \Phi\left(\frac{a-m}{\sigma}\right)$.

For $a = -b$ it takes on the widely known form:

$$P\{-b \leq x < b\} = \Phi\left(\frac{b+m}{\sigma}\right) + \Phi\left(\frac{b-m}{\sigma}\right).$$

The normal distribution is very useful for obtaining integral estimates. For example, the confidence limits corresponding to probability P, for a normal distribution of a random variable with mean m and SD σ are calculated from the formulas:

$$- \quad \Delta_P = m - \lambda(P)\sigma,$$

$$\Delta_P = m + \lambda(P)\sigma,$$

where $\lambda(P)$ is the two-sided quantile of the normal distribution, corresponding to probability P, and calculated from the formula $\lambda(P) = \Phi^{-1}(2P - 1)$.

2.2.3   Generalized Normal Distribution Law

The validity of the third postulate, formulated by Gauss in his development of the normal distribution law, just like many other axioms that lie at the basis of mathematical theories, can neither be proven theoretically nor verified experimentally. Hence, for two centuries, it has been subject to some doubt. In particular, it has been proven that for many measuring instruments it is not fulfilled. Let us introduce an example of such an instrument.

Example 2.1. The equation of electrodynamic measurements has the form

$$WY^2 = K\frac{X^2}{R^2} + M,$$

where X is the measurand (input instrument signal), Y is the reading (output instrument signal), R is input resistance, W is spring tension, K is an electrodynamic constant, and M is the moment of friction in the supports, the sign of which is determined by the direction of change of the input signal.

The main source of random measurement error is friction in the supports, subject to the normal law. To reduce this error, a series of $2n$ measurements are conducted, each time reversing the direction of the voltage changing, and the measurement result is determined from the formula for the mean:

$$\bar{X} = R\sqrt{\frac{W}{K}}\frac{1}{2n}\sum_{i=1}^{n}(Y_{+,i} + Y_{-,i}),$$

where $Y_{+,i}$, $Y_{-,i}$ are instrument readings, as the input signal increases or decreases. However, this estimation of the measurand is not the best one. Actually, it is a biased estimator, since the moment of friction in the supports and, consequently, the error due to friction is not totally excluded.[1] To the contrary, the computation of the mean,

$$\bar{X}_2 = R\sqrt{\frac{W}{K}\frac{1}{2n}\sum_{i=1}^{n}(Y_{+,i}^2 + Y_{-,i}^2)},$$

makes it possible to fully exclude the error from friction. One may show that $\bar{X}_2$ is an unbiased, consistent, and efficient estimate of this value; i.e., its best valuation.

The example introduced shows that the mean of measurement results is not always the best valuation of the measurand. A theoretically proven and stronger assertion is that the mean is an efficient estimator of the measurand when the measurement errors are normally distributed. Hence, if the distribution rule differs from the normal, finding its mean is not the best solution.

Nevertheless, one ought not bring into doubt the merit of the wide utilization of the normal distribution in the statistical processing of a series of experiments. Processing the results of measurements must be concluded by determining the interval in which the measurand lies. And if the methods of statistical modeling are not applied, this is practically feasible only with a normal distribution of a series of measurements, since only this fully ensures the statistical distributions necessary for solving this problem. Hence, in practice, processing of measurement results proceeds as a rule by a presentation regarding its normal distribution. For this reason, what is pressing is also the generalization of the normal distribution of random measurement errors using another probability law which, while preserving the advantages of a normal distribution, at the same time would be, due to its flexibility, useful in more precisely approximating the sample distributions of a series of tests. It is possible to derive such a distribution if, in Gauss' axiomatic development in the former section, we replace the third postulate with the following:

If any quantity is determined from many equally precise measurements $x_1, \ldots, x_n$, *then the power mean* $u = \left(1/n\sum_{i=1}^{n}x_i^F\right)^{1/F}$ *of all observed values with*

---

[1] Some information from the theory of statistical valuations:

- *A statistical valuation of a quantity is the best if it is unbiased, consistent, and efficient;*
- *a statistical valuation is called consistent if it approaches the true value of the quantity as the amount of experimental data increases;*
- *a statistical valuation is called unbiased if its mean is equal to the measurand;*
- *a statistical valuation is called efficient if its SD is less than the SD of any other estimate of this quantity.*

parameter F, the value of which is defined along the series x_1, ..., x_n, will be the most probable value.

In this regard, the arbitrary exponent F is determined from the sample and can be equated to 1 in a particular case. This approach was first employed in 1955 by I.G. Fridlender [14]. As a result, he derived a new distribution that generalizes the normal law. However, this distribution had a significant deficiency: by limiting the region of dispersion to non-negative values of the quantities, then contradicting the first postulate of Carl Gauss as well as the nature of random errors. But this deficiency can be easily removed if one takes as the true value of the measurand for $F \neq 0$ the limit as $n \to \infty$ of the value $\hat{x} = \text{sign}[\hat{z}] \times [|\hat{z}|]^{1/F}$, where $\hat{z} = 1/n \sum_{i=1}^{n} \text{sign}(x_i)(|x_i|)^F$ (sign(y) is the sign of the quantity y) [15]. In the particular case for $F = 0$, one may take as the true value the limit as $n \to \infty$ of the value $\hat{x} = \text{sign}[\hat{z}_i] \times \exp[\hat{z}_i]$, where $\hat{z}_i = 1/n \sum_{i=1}^{n} \text{sign}(x_i) \times \ln(|x_i|)$.

From Gauss' axiomatic development in this case, it follows that by the normal law these values will be distributed:

$$z_i = \text{sign}(x_i) \times \begin{cases} (|x_i|)^F, & F \neq 0, \\ \ln(|x_i|), & F = 0. \end{cases} \tag{2.19}$$

Since the values z_i are subject to normal distribution with density $f(z) = \frac{1}{\sqrt{2\pi}} e^{-(z-m_F)^2/2\sigma_F^2}$, mean m_F, and standard deviation σ_F, the probability distribution of the values of x_i is equal to [15]

$$f(x) = \frac{\partial z/\partial x}{\sqrt{2\pi}\sigma} e^{-(z-m_F)^2/2\sigma_F^2} = \begin{cases} \frac{|F| \times (|x|)^{F-1}}{\sqrt{2\pi}\sigma} e^{-(\text{sign}(x)(|x|)^F - m_F)^2/2\sigma_F^2}, & F \neq 0, \\ \frac{1}{\sqrt{2\pi}\sigma \times |x|} e^{-(\text{sign}(x)\ln(|x|) - m_{\ln})^2/2\sigma_{\ln}^2}, & F = 0. \end{cases} \tag{2.20}$$

Expression (2.20) describes the density of a generalized normal distribution of random measurement error. In contrast to the normal distribution, this distribution is triparametric: to parameters m_F and σ_F is added F. Here, the valuations of m_F and σ_F depend on F. Figure 7 shows graphs of this distribution's density for various values of m_F, σ_F, and F. They demonstrate that by varying parameter F, it is possible to derive distribution densities that differ from each other in principle: symmetric and nonsymmetric, gently sloping top and sharp-peaked, unimodal and bimodal.

The graphs of the coefficient of skewness and kurtosis, shown in Fig. 8, also substantiate this. It is know that the coefficient of skewness of a normal distribution is zero and of kurtosis is three. The graphs show that by varying F, it is possible to obtain a coefficient of skewness in the range from 0 to 5 and kurtosis from 1 to 20.

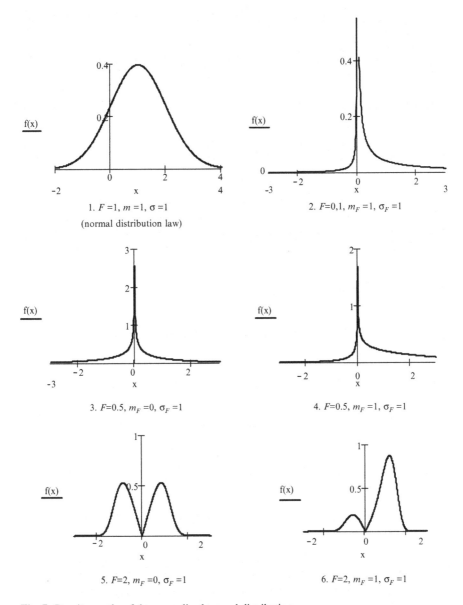

Fig. 7 Density graphs of the generalized normal distribution

Hence by selecting a value for F, it is possible to approximate any experimental series with a high degree of precision using a generalized normal distribution. Here, the possibility is preserved to derive the statistical limits for the parameters of this distribution, since the values of z_i, defined by formula (2.19), are subject to the normal distribution.

Fig. 8 Graphs of the
coefficient of skewness
and coefficient of kurtosis
as functions of F.
(1) Kurtosis for
$k = m_F/\sigma_F = 0$.
(2) Coefficient of skewness
for $k = m_F/\sigma_F = 1$

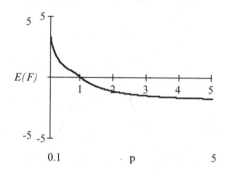

The probability that a random measurement error subject to the generalized normal distribution is located in the interval (a, b), is computed according to this formula, analogous to (2.18):

$$P\{a<x\leq b\} = \int_a^b f(x)\,\mathrm{d}x$$

$$= F\left(\frac{\mathrm{sign}(b) \times h(b) - m_F}{\sigma_F}\right) - F\left(\frac{\mathrm{sign}(a) \times h(a) - m_F}{\sigma_F}\right), \quad (2.21)$$

in which

$$h(x) = \begin{cases} (|x|)^F, & F \neq 0, \\ \ln(|x|), & F = 0, \end{cases}$$

is the function at $x = a$ or b, and m_F, σ_F are as in (2.20).

2.2.4 Basic Statistical Distributions Applied in Processing Measurement Results

The hypothesis of the correspondence of the distributions of random measurement errors to the normal law has facilitated the establishment of a number of probability distributions of random values that are widely used in statistical processing of measurement results. The following distributions are the ones most commonly used.

The χ^2 distribution [16]

Let $x_1, x_2, ..., x_n$ be normally distributed random values with mean m and SD σ. After the replacement of variables $\xi_i = (x_i - m)/\sigma$, the series of normally distributed random values $\xi_1, \xi_2, ..., \xi_n$ will have a mean of zero and an SD equal to 1. The distribution function

$$\chi^2 = \sum_{i=1}^{n} \xi_i^2$$

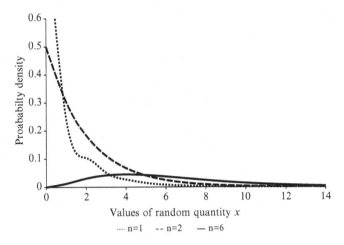

Fig. 9 χ^2– distribution density

is called the χ^2 distribution (chi-square distribution). This distribution plays an important role in metrology. The density of the χ^2 distribution has the form

$$\phi(x) = \frac{x^{(n/2)-1}e^{-x/2}}{2^{n/2}\Gamma(n/2)}, \quad x \geq 0, \tag{2.22}$$

where $\Gamma(x)$ is the gamma function, defined by the equation $\Gamma(x+1) = (x+1)\Gamma(x)$ (which for positive integers x satisfies the equality $\Gamma(x) = x!$), n is a distribution parameter, referred to as the degree of freedom.

Figure 9 shows a graph of this probability density for $n = 1, 2,$ and 6.

The most well-known application of the χ^2 distribution is in verifying the hypothesis regarding the form of the distribution laws to be used for the measurement results.

Student's distribution [16]

Student's distribution describes a probability density for a mean that is calculated by sampling from n random measurement results of one and the same quantity, distributed according to the normal law. Student's distribution is derived as follows. A series $x_1, x_2, \ldots, x_n$ of normally distributed random values with mean m and SD σ is examined. After replacement of variables $\xi_i = x_i - m/\sigma$, the series of normally distributed random values $\xi_1, \xi_2, \ldots, \xi_n$ will have a mean of zero and an SD equal to 1.

By force of the reproducibility of the normal distribution, the mean of this sample $\xi = 1/n \sum_{i=1}^{n} \xi_i$ is likewise distributed by the normal law with density $f(\xi) = \sqrt{n/2\pi}e^{-0.5n\xi^2}$. The sampling standard deviation η of the series ξ_1, $\xi_2, \ldots, \xi_n$ is defined by the formula $\eta = \sqrt{1/n \sum_{i=1}^{n} \xi_i^2} = \chi/\sqrt{n}$, where χ is as in

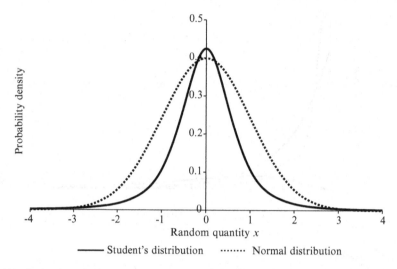

Fig. 10 Student's probability density for $n = 3$

(2.21). W.S. Gosset proved that the quotient from dividing these independent random values, $\zeta = \xi/\eta$, has the probability density

$$f(\zeta) = \frac{\Gamma((n+1)/2)}{\sqrt{\pi}\Gamma(n/2)}(1 + \zeta^2)^{-(n+1/2)}. \qquad (2.23)$$

Formula (2.23) describes a family of Student's (Gosset's pseudonym) distributions, as a function of the number n of degrees of freedom. Student's distribution is symmetrical about zero and thus its mean is equal to zero. The dispersion of this distribution is equal to $D(\zeta) = n/(n - 2)$. As n increases, Student's distribution transitions to the normal distribution, although for small n it noticeably differs from it. Figure 10 shows a graph of the Student's probability density for $n = 3$. The dotted line shows the normal distribution density for $m = 0$, $\sigma = 1$.

Student's distribution is widely used in processing results of multiple measurements. Since the mean $\bar{x}$ of a normally distributed sample is subject to Student's distribution, the confidence interval for it is calculated using the formula

$$\left(\bar{x} - t(n-1, P)\frac{S}{\sqrt{n}};\ \bar{x} + t(n-1, P)\frac{S}{\sqrt{n}}\right), \qquad (2.24)$$

where $\bar{x} = 1/n\sum_{i=1}^{n}x_i$ is the result of multiple measurements, $S = \sqrt{1/(n-1)\sum_{i=1}^{n}(x_i - \bar{x})^2}$ is the sample SD of the measurement results,

$t(n - 1, P)$ is the Student's distribution quantile with number $n - 1$ of degrees of freedom, and confidence probability P.

Fisher's distribution [16]

Let $\xi_1, \xi_2, ..., \xi_m, \eta_1, \eta_2, ..., \eta_n$ be normally distributed independent random values with parameters 0 and σ. As shown above, the values $\xi = 1/\sigma^2 \sum_{i=1}^{m} \xi_i^2$ and $\eta = 1/\sigma^2 \sum_{i=1}^{n} \eta_i^2$ have a χ^2-distribution with m and n degrees of freedom. Then, as R.A. Fisher demonstrated, the value

$$\tau = \frac{\xi}{\eta} = \frac{\sum_{i=1}^{m} \xi_i^2}{\sum_{i=1}^{n} \eta_i^2},$$

has the probability density:

$$f_{mn}(\tau) = \frac{\Gamma((m+n)/2)}{\Gamma(m/2)\Gamma(n/2)} \times \frac{\tau^{(m/2)-1}}{(\tau + 1)^{(m+n)/2}}, \quad x > 0. \qquad (2.25)$$

The basic use of Fisher's distribution is to verify the hypothesis regarding equality of the dispersions of two series of measurements.

2.3 Systematic Measurement Errors

Systematic errors distort measurement results most substantially. Hence great significance is assigned to the detection and exclusion of systematic errors. Systematic errors are differentiated according to their source of origin, as caused by

- The properties of the measurement facilities;
- Deviation of the measurement conditions from normal conditions;
- Imperfection in the method of measurement;
- Error in operator actions.

Let us examine these components of systematic measurement error.

2.3.1 Systematic Errors Due to the Properties of Measurement Equipment and the Measurement Conditions Deviating from Normal Conditions

The sum of these errors is often called measurement *instrumental error*. As a rule, instrumental error brings a basic contribution to the error in measurement results. In formalized form, the reasons for the occurrence of instrumental error can be presented as follows. Let

$$y = f(x, R_i, \xi_j, P, \tau), \quad i = 1, ..., n, \quad j = 1, ..., m \tag{2.26}$$

be a real function of the transformation of the measuring instrument at the sampling instant, expressing the dependence of the output signal of the measuring installation y on the measured value x, parameters R_i of the components of the measuring instrument, the conditions of measurement, and the inconclusive parameters of the measurement signal[2] ξ_j, energy P extracted by the measuring instrument from the object of measurement, and the delay time of the measurement signal τ. The result of measurements, equal to

$$\hat{x} = f_K^{-1}(y), \tag{2.27}$$

is determined from the calibrating function $y = f_K(x)$, assigned to the measuring instrument at its last calibration. It will not be burdened with the systematic component of instrumental error if the following conditions are satisfied:

$$f_K(x) = f_{nom}(x); \tag{2.28}$$

where $y = f_{nom}(x)$ is a *nominal transform function of the measuring instrument, equal to the theoretical dependence of the transform function of the measurement signal in the measuring instrument $y(x)$ in accordance with the measurement method implemented*;

- the values of the parameters R_i of the components of the measuring instrument at the time of measurement will precisely coincide with their values $R_{i.K}$ at the time of the last calibration of the measuring instrument;
- $\xi_j = \xi_{jnom}$, $j = 1, ..., n$, the measurement conditions, coincide with the normal conditions and the non-information parameters of the input signal are equal to zero;
- $P = 0$, the energy extracted by the measuring instrument from the object of measurement, is equal to zero;
- $\tau = 0$, the time delay, is absent.

As a consequence of these conditions being fulfilled, the following equality will be valid:

$$f(x, R_i, \xi_j, P, \tau) = f(x, R_{i.k}, \xi_{jnom}, 0, 0) = f_K(x). \tag{2.29}$$

Let us substitute expression (2.28) into (2.27):

$$\hat{x} = f_k^{-1}[f(x, R_i, \xi_j, P, \tau)]. \tag{2.30}$$

We expand this function into a Taylor series in the neighborhood of the straight line $(x, R_{i.k}, \xi_{jnom}, 0, 0)$. Here, since the systematic error is a small value compared with the measurement results, it can be bounded by the first derivatives of this series. Let us designate as $\partial f(x, R_{i.k}, \xi_{jnom}, 0, 0)$ the derivative of

$\partial f(x, R_i, \xi_j, P, \tau)$ for $R_i = R_{i.k}$, $\xi_j = \xi_{jnom}$, $P = 0$, $\tau = 0$. From expressions (2.28) and (2.29) we derive:

$$f_K^{-1}[f(x, R_{i,k}, \xi_{jnom}, 0, 0)] = f_{nom}^{-1}[f(x, R_{i,k}, \xi_{jnom}, 0, 0)] = x,$$

$$\frac{\partial f_K^{-1}[f(x, R_{i,k}, \xi_{jnom}, 0, 0)]}{\partial f(x, R_{i,k}, \xi_{jnom}, 0, 0)} = \frac{\partial x}{\partial y} = \frac{1}{\partial y / \partial x} = \frac{1}{W(x)},$$

where $W(x)$ is the function of the sensitivity of the measuring instrument to the input signal.

Then formula (2.30) can be written as follows:

$$\hat{x} = x + \frac{1}{W(x)} \left[\Delta f + \sum_{j=1}^{m} W(\xi_j) \times (\xi_j - \xi_{jnom}) + W(P) \times P + W(\tau) \times \tau \right],$$

where $\Delta f = [f(x, R_i, \xi_{jnom}, 0, 0) - f_{nom}(x)] = \Delta f_1 + \Delta f_2$, $\Delta f_1 = [f_k(x) - f_{nom}(x)]$ is the deviation of the calibration function from the nominal transform function of the measuring instrument, $\Delta f_2 = [f(x, R_i, \xi_{jnom}, 0, 0) - f_k(x)] = \sum_{i=1}^{n} W(R_i)$ $(R_i - R_{ik})$ is the deviation of the actual transform function from the calibrated dependency, due to instability of the measuring instrument components,

$$W(\xi_j) = \frac{\partial f(x, R_{i.k}, \xi_{jnom}, 0, 0)}{\partial \xi_j}$$

is the function of the sensitivity of the measuring instrument to the jth measurement condition (or to a non-information parameter of the input signal),

$$W(P) = \frac{\partial f(x, R_{i.k}, \xi_{jnom}, 0, 0)}{\partial P}$$

is the function of the sensitivity of the measuring instrument to the energy extracted from the object of measurement,

$$W(R_i) = \frac{\partial f(x, R_{i.k}, \xi_{jnom}, 0, 0)}{\partial R_i}$$

is the function of the sensitivity of the measuring instrument to a change in the parameter R_i of the components of the measuring instrument, $W(\tau)$ is the function of the sensitivity of the measuring instrument to signal time delay, which is equal to

$$W(\tau) = \frac{\partial f(x, R_{i.k}, \xi_{jnom}, 0, 0)}{\partial \tau} = \frac{\partial f(x, R_{i.k}, \xi_{jnom}, 0, 0)}{\partial x} \cdot \frac{\partial x}{\partial \tau} = W(x) \cdot \frac{\partial x}{\partial \tau},$$

since

$$\frac{\partial f(x, R_{i.k}, \xi_{jnom}, 0, 0)}{\partial x} = \frac{\partial y}{\partial x} = W(x),$$

and $\partial x/\partial \tau$ is the rate of change of the sensing signal.

Consequently, the instrumental component of systematic measurement error, $\Delta_{\text{instr.}} = \Delta x = \hat{x} - x$, is equal to

$$\Delta_{\text{instr.}} = \frac{1}{W(x)} \left[\Delta f_1 + \Delta f_2 + \sum_{j=1}^{m} W(\xi_j) \times (\xi_j - \xi_{jnom}) + W(P) \times P \right] + \frac{\partial x}{\partial \tau} \tau.$$

(2.31)

The formula presents the basic groups of the constituents of instrumental measurement error. The first two members characterize the first group, called the intrinsic error. *Fundamental measurement error is measurement error under normal measurement conditions.* It arises as a consequence of the deviation of the actual transform function of the measuring instrument from the nominal transform function. $\Delta f_1/W \times (x)$ is that part of the intrinsic error caused by the difference between the calibration function and the nominal transform function, which reflects the transformation of the measurand in precise correspondence with the method of its measurement. Primarily, it is the consequence of measurement error when calibrating the measuring instrument. In addition, it is caused by imperfection in the design of the measuring instrument and the technology for its manufacture. $\Delta f_2/W(x)$ is the second part of intrinsic error, caused by the difference of the actual transform function of the measuring instrument at the sampling instant from the calibrated function. This is due to instability of the measuring instrument, caused by fatigue and wear of measuring instrument components and the accumulation of various fault conditions (deformation, corrosion, and so forth), and small defects due to mechanical, thermal, or electrical overloads.

The third term characterizes errors of the second group, referred to as supplemental errors. They contain the errors $\Delta_{\partial j}$, caused by the sensitivity of the measuring instrument to changes in the j influencing quantities and the non-information parameters of the input signal relative to their nominal values. These can be thermal and air currents, magnetic and electrical fields, changes in atmospheric pressure, air humidity, and vibration. The ambient temperature can significantly distort the measurement results, especially with the nonuniform effect on the measuring instrument or the object of measurement. Magnetic fields created near positioned electrical devices, transformers, and wires cause magnetization of any moving elements of a measuring instrument that are made of magnetic materials, and thus their mutual attraction and deviation from normal position. Errors occur also as the result of the effect of electrical fields. The effect of magnetic and electrical fields on the accuracy of measurements increases with higher frequency of an AC current that creates this field. Temperatures of phase transitions (boiling point, solidifying, and melting) of various pure substances and compounds widely used in temperature and analytical measurements depend significantly on atmospheric pressure. Hence in these types of measurements, error in determining such pressure is also a source of systematic error. Moisture can also be a reason for supplemental errors. Moisture in an object of measurement (such as petroleum and natural gas) is a non-information parameter of the measuring signal that distorts the measurements results (such as net weight of petroleum and the heat value of natural gas when measured by the chromatographic method). Moisture of the ambient air affects hydroscopic materials, modifying their properties (such as electrical resistance).

The fourth term characterizes the third group of errors – errors that are formed as a result of the interaction of the measuring instrument and the object of measurement. We shall see the essence of these errors in the following example.

Example 2.2. Measuring an electrical resistance R by comparison with a known resistance R_0 can be done by comparing the currents crossing these resistances with successive connection to a DC current source. The measurement equation, without accounting for ammeter resistance, is $R = R_0(I_0/I)$ (I and I_0 are the currents when connecting R and R_0), while the actual measurement equation is $R = (R_0 + r)(I_0/I) - r$, where r is the resistance of the ammeter. The relative measurement error caused by the interaction of the ammeter with the object of measurement is $\delta R = r/R \times (I_0 - I/I)$. Consequently, this method of measurement can only be used to measure large ($R >> r$) resistances.

The fifth term characterizes the fourth group of errors – errors caused by inertia of the measuring instrument and the rate of change of the input signal. These are called dynamic errors.

2.3.2 Normalizing Metrological Characteristics of Measuring Equipment

In developing measurement methodology, one must select a measuring instrument that will guarantee the necessary measurement accuracy. However, as follows from the preceding section, the special feature of all enumerated groups of errors, except for the first group, consists in the fact that they are associated not only with the properties of the measuring instrument but also with the measurement conditions. Hence in the process of developing this methodology, one must evaluate the instrumental component of measurement errors in the specified measurement conditions. In connection with this, when any measuring instrument is being developed, the technical characteristics of a special type, referred to as the metrological characteristics, are standardized and specified (*the properties of the measuring instrument that affect measurement error are called the metrological properties, and the characteristics of these properties are the metrological characteristics of the measuring instrument*). The system of notations of the metrological characteristics of a measuring instrument and the methods for standardizing them are established in [17]. The metrology of standardization, established by this standard, proceeds from the following.

Normalizing metrological characteristics are essential in solving two basic problems:

- monitoring each model of a measuring instrument for compliance with established standards,
- determining measurement results and the a priori valuation of measurement instrumental error.

Here out must keep in mind that the metrological properties of each specific model of a measuring instrument are constant at a specific moment in time, but in the aggregate of measuring instruments of a given type, they vary in a random manner. This occurs as a consequence of the scattering of manufacturing parameters when building the measuring instrument, and the differentiation of the conditions of use resulting in the random nature of the processes of wear and aging of its components and the random measurement error with periodic calibrations of the measuring instrument, and other similar reasons. Hence two types of normalizing metrological characteristics are theoretically possible. Limits of allowable values for this type of metrological characteristics of a measuring instrument pertain to characteristics of the first type, almost exclusively used in practice. They are used both to monitor the suitability of each model of a measuring instrument and to evaluate the maximally possible instrumental measurement error. Characteristics of the second type, used extremely rarely, relate to the mean and SD of the values of a measurement characteristic; these are computed in the aggregate for measuring instruments of this type and are convenient for the evaluation of the instrumental measurement error by the statistical summation method.

Thus, the characteristics of the systematic component of the basic error Δx_c are either the limits of its allowable values $\pm \Delta_c$ or else its limits and the mean m_c and SD σ_c, wherein it is permissible to use the second method of normalization if it is possible to ignore changes in these characteristics under extended use and in various measurement conditions. In other cases, only the $\pm \Delta_c$ are normalized. For many measuring instruments in which several systematic components of basic error are differentiated, one may normalize the limits of allowable values of these components, $\pm \Delta_{ci}$, in place of the $\pm \Delta_c$. For this, the condition must be satisfied:

$$\sqrt{\sum_{i=1}^{m} k_i^2 \Delta_{ci}^2} \leq \Delta_c. \tag{2.32}$$

Example 2.3. For many measuring instruments, some limits are normalized instead of limits on systematic error $\pm \Delta_c$: absolute additive error $\pm \Delta_a$, relative multiplicative error $\pm \delta_m$, and the adjusted error (to the maximum value x_{max} of the range of the measuring instrument), caused by the non-linearity of the calibration function $\pm \delta_{nl}$. The values of these limits must satisfy this condition:

$$\sqrt{\Delta_a^2 + (\delta_m x)^2 + (\delta_{nl} x_{max})^2} \leq \Delta_c(x),$$

where x is the value of the measurand and $\Delta_c(x)$ is the limit of absolute systematic error at this point of the range of the measuring instrument.

The characteristics of this group also relate to characteristics of the random error $\dot{\Delta}$, the limit of allowable values of the standard deviation of the random error $\sigma\dot{\Delta}$ and to the characteristics of the random error from hysteresis $\dot{\Delta}_H$, the limit H of the allowable variation of the input signal to the measuring instrument.[2]

In those cases when the random error of the measuring instrument is insignificant, it is recommended to normalize the characteristics of the error of the measuring instrument – the limits $\pm\ \Delta$ of the allowable error of the measuring instrument and the limit H of the allowable variation in the input signal to the measuring instrument. It is possible also to normalize one generalized characteristic of the intrinsic error of measurement – the limit $\pm\Delta_o$ of the allowable intrinsic error of the measuring instrument.

The characteristics of the sensitivity of the measuring instrument to influencing quantities and to the non-information parameters of the sensing signal ξ_j are source functions of $\Psi(\xi_j) = W(\xi_j)/W(x)$. In normalization, the nominal source function of $\Psi_{nom}(\xi_j)$ and the limits $\Delta\Psi(\xi_j)$ of allowable deviations from it are established. The nominal source functions serve to determine corrections for systematic errors caused by the difference between the values of the influencing quantities and their nominal values. The limits $\Delta\Psi(\xi_j)$ are used to monitor the quality of the measuring instrument and evaluate the residual systematic error left after introducing the corrections. If the measuring instrument of one type has great scattering of the source function (i.e., $\Delta\Psi(\xi_j)>0.2\Psi_{nom}(\xi_j)$), the determination of corrections taking account of the $\Psi_{nom}(\xi_j)$ can introduce a significant error into the measurement results. Hence for specific samples of such measuring instruments, it is advisable to show the individual source functions used to determine corrections and to normalize the boundary source functions of this type to monitor quality and evaluate the residual of systematic error: $\Psi^+(\xi_j) = \Psi_{nom}(\xi_j) + \Delta\Psi(\xi_j)$ and $\Psi^-(\xi_j) = \Psi_{nom}(\xi_j) - \Delta\Psi(\xi_j)$.

It is permissible to normalize the sensitivity of the measuring instrument to influencing quantities by another method: by fixing the limits $\varepsilon^-(\xi_j)$, $\varepsilon^+(\xi_j)$ of allowable changes to the metrological characteristics of the measuring instrument that are caused by changes in the influencing quantities within established limits.

Precisely the same way, by establishing the nominal characteristics and the limits of allowable deviations from them, the sensitivity characteristics of the measurement results to the energy extracted from the object of measurement by the measuring instrument $\Psi(P) = W(P)/W(x)$, as well as the dynamic characteristics of the measuring instrument as recommended by [16], are normalized: the transitional characteristics h, amplitude–phase characteristics

---

[2] *Two Variation of the input signal is the name given to the difference between the two means of the informational parameter of the input signal of the measuring instrument, derived during measurements of a quantity that has one and the same value, with a smooth, slowly varying approach to this number from the upper and lower sides.*

$G(j\omega)$, amplitude–frequency characteristics $A(\omega)$, reaction time t_r, and others. For measuring instruments which have large (more than 20% from the nominal characteristics) scatter of these characteristics in the aggregate of measuring instruments of this type, the boundary characteristics used to monitor quality and evaluate the residuals of the systematic measurement errors are normalized, and the corrections to the measurement results are determined with the aid of the individual characteristics of each sample measuring instrument. For measuring instruments for which this scatter is less than 20%, the nominal characteristics and limits of allowable deviations from them are normalized.

Example 2.4. Calculating the instrumental measurement error with an analog voltmeter [18]

1. Input data

   (a) Measured voltage $W = 0.6$ V.
   (b) Normalized metrological characteristics of the measuring instrument:

   – limit of allowable intrinsic error $\Delta_0 = 20$ mV,
   – boundary source function of the temperature ξ_1 on the error $\Psi(\xi_1) = 0.5$ mV/°C,
   – limit of permissible changes in error due to deviation of the voltage from the nominal value ($\xi_{2nom} = 220$ V) by $\pm 10\%$, is $\varepsilon^+(\xi_2) = 10$ mV,
   – nominal amplitude–frequency characteristics[3]

   $$A(\omega) = \frac{A(\omega_0)}{\sqrt{1 + \omega^2 T^2}},$$

   where $T = 5$ ms is a time constant, $\omega_0 = 0$, $A(\omega_0) = 1$.

   (c) Characteristics of influencing quantities:

   $$\xi_{1nom} = 20°C, \quad \xi_1^- = 25°C, \quad \xi_1^+ = 35°C, \quad \xi_2^- = 200B, \quad \xi_2^- = 230B,$$
   $$\omega = (0 - 10) \ \Gamma \amalg .$$

2. Calculation of the greatest possible values of supplemental errors

$$\Delta_{\partial 1}^+ = \psi(\xi_1)(\xi_1^+ - \xi_{1nom}) = 0.5 \, (35 - 20) = 7.5 \ mV, \quad \Delta_{\partial 2}^+ = \varepsilon^+(\xi_2) = 10 \ mV.$$

3. Top-down analysis of the relative dynamic error of a linear measuring instrument is calculated with the formula

---

[3] *The amplitude–frequency characteristic is the ratio, depending on the angular frequency ω, of the amplitude of the output signal of a linear measuring instrument to the amplitude of an input sinusoidal signal in steady-state mode [18].*

$$\delta_{dyn}^+ = \left| 1 - \frac{A(\omega_0)}{A(\omega)} \right|$$

[18]. Consequently,

$$\delta_{dyn}^+ = \left| 1 - \frac{1}{A(\omega)} \right| = \left| 1 - \sqrt{1 + \omega^2 T^2} \right| = \left| 1 - \sqrt{1 + (10 \times 0.005)^2} \right| \cong 0.025.$$

4. Estimation of the maximum instrumental error in specified conditions of use:

$$\Delta_{instr} = \pm(\Delta_o + \Delta_{\partial 1} + \Delta_{\partial 2} + \delta_{dyn}W) = \pm(20 + 7.5 + 10 + 0.025 \times 600)$$
$$= \pm 52.5 \; MB.$$

This estimation was derived, in accordance with [18], by arithmetic summation of components. If one uses the mean-square summation, as recommended by the international Guide [19], then

$$\Delta_{instr} = \pm\sqrt{\Delta_0^2 + \Delta_{\partial 1}^2 + \Delta_{\partial 2}^2 + (\delta_{dyn}W)^2}$$
$$= \pm\sqrt{20^2 + 7.5^2 + 10^2 + (0.025 \times 600)^2} = \pm 28.0 \; mV.$$

2.3.3 Measurement Method Errors

These errors can occur due to imperfection in the selected measurement method, due to the limited accuracy of empirical formulas used in to describe the phenomenon positioned as the basis of the measurement, or due to limited precision of the physical constants used in the equations. One must also include here errors caused by the mismatch between the measurement model adopted and the actual object, due to assumptions or simplifications that were taken. In some cases, the effect of these assumptions on measurement error turns out to be insignificant, but in others it can be substantial. An example of an error caused by oversimplification of the measurement method is ignoring the mass of air compressed, according to Archimedes' law, using a balance weight or hanging on beam scales. In conducting working measurements, this is usually ignored. However, in precise measurements, one needs to consider and introduce an appropriate correction. Another example is measuring the volumes of bodies whose form is taken to be (in the measurement model) geometrically straight, by taking an insufficient number of linear measurements. Hence, a substantial methodological error can result from measuring one length, one width, and one height. For more accurate measurement, one should measure these parameters along each face at several positions.

Method errors are inherent in measurement methods that are based on test data that lack strong theoretical foundation. An example of such methods would be the various methods of measuring the hardness of metals. One of them (the Rockwell method) determines hardness by the submerged depth, in the tested metal, of the tip of a specified form under the effect of a specified force impulse. The foundation of other methods (Brinnel and Vickers) is the relationship between the hardness and the size of an impression left by a tip under specified conditions of action. Each of these methods determines hardness using its own scales, and conversion of the measurement results of one into another is only approximate. This is explained by the fact that the specified methods use different phenomena that purportedly characterize hardness.

Estimates of error in formulas and physical constants are mostly known. When they are unknown, errors in the empirical formulas are transferred into a series of random values, using the process of randomization. For this purpose, the same quantity is measured with several methods, and the test data derived is calculated using its mean square value.

Analytical measurements differ from the others in the fact that they incorporate a series of preparatory operations: the selection of a sample of the analyzed object, its delivery to the measuring laboratory, storage, preparation of the sample for instrumental operations (cleaning, drying, transition to another phase state, etc.), preparation of calibration solutions, and other. These operations are often not considered with regard to the accuracy characteristics of the measurement method, when considering the measurement simply as its instrumental component. It is easy to prove that this position is erroneous. Let us recall that a measurement error is the deviation of a measurement result from the true value of a measurand. Let us suppose that it is essential to estimate some quantity that reflects a physical and chemical property of an object (for example, the density of a product in a batch, or the concentration of a chemical component in lake water or soil at a settlement). The true value of this quantity, and not a sample taken from it, must characterize this object. The user of measurement information is interested specifically in this, and if there has been distortion of the measurement results, he does not care at what stage it was introduced. Consequently, error in analytical measurement must also account for errors in preparatory operations.

The necessity of accounting for such operations is due to the fact that the risk of introducing systematic errors into the measurement results in these operations is incomparably higher than in instrumental. In practice, systematic measurement error can occur in these operations due to the effect of many possible sources, in particular:

– that the sample extracted from the object may not be representative (not adequately representing the measurand),
– that the sample being measured may have changed since the time that the sample was taken,
– the effect of non-information parameters (disturbing the sample components),

- contamination of the sampling unit and laboratory vessels used in preparing the sample,
- inaccurate measurement of environmental parameters,
- errors in measuring masses and volumes,
- errors in preparing calibration solutions [20].

2.3.4   Systematic Errors Due to Operator's Inaccurate Performance

These errors, referred to as subjective, are as a rule the consequence of the person's individual traits, caused by his organic properties or by deep-rooted bad habits. For example, an operator's invalid actions may lead to a delay in recording a measuring signal or to an asymmetry in setting an indicator between the guide lines. Reaction time to a received signal plays an important role in the occurrence of subjective systematic errors. This is different for different people, but is relatively stable for each person over a more or less extended period. For example, a person's reaction speed to a light signal varies between 0.15 and 0.225 s and to an auditory signal between 0.08 and 0.2 s. The famous astronomer Bessel compared the accuracy of the measurements of time during star transit taken by various astronomers and himself. He established that there were large discrepancies, although stable, between his data and that of other researchers. Bessel came to the conclusion that the reason for these systematic errors lay in the different reaction speeds of each of the astronomers [3].

Currently, in connection with automated recording of measurement information, which is imposed by the demand for high accuracy, subjective measurement errors have lost their significance.

2.3.5   Elimination of Systematic Errors

Systematic errors introduce a shift into measurement results. The greatest danger is in unrecognized systematic errors whose existence is not even suspected. It is systematic errors that more than once have been the reason for erroneous scientific conclusions, production breakdown, and irrational economic losses. Hence systematic errors must be removed as much as possible by some method. Methods for removing systematic errors can be divided into the following groups:

- removing the sources of error before commencing the measurements (prophylaxis);
- excluding systematic errors during the measurement process;
- introducing known corrections to the measurement results.

The first method is the most rational since it significantly simplifies and speeds up the measurement process. The term 'removal of a source of error' is understood to be both its removal (such as removing a heat source) and protection of the object of measurement and the measurement apparatus from the effect of such sources. To prevent the appearance of temperature errors, thermostatic control is used – stabilization of the ambient temperature in a narrow range of values. Both the rooms where the measurements take place, and the measuring instruments overall and their constituent parts, are thermostatically controlled. Protecting the measuring instrument from the effect of the Earth's magnetic field and from magnetic fields induced by DC and AC circuits is done with magnetic shields. Harmful vibrations are eliminated by cushioning (dampening vibrations of) the measuring instrument. Sources of instrumental error that are inherent in the specific instance of a measuring instrument can be eliminated, before measurements begin, by conducting a calibration. Likewise, sources of error associated with improper installation of the measurement unit can be eliminated before measurement commences.

During measurements, some instrumental errors can be excluded, being errors from improper setup and errors from disruptive influences. This will be achieved by using a number of specialized approaches associated with repeated measurements. These are the methods of replacement and contraposition. In the replacement method, the quantity being sought is measured, and with repeated measurement the object of measurement is replaced with a measure located in the same conditions as itself. Determining the measurement result from the value of this measure, exclusions are added for the large number of systematic effects that affect the equilibrium position of the measurement layout. For example, when measuring the parameters of an electrical circuit (electrical resistance, capacitance, or inductance), the object is connected to a measurement circuit and put into equilibrium. After equilibrium is reached, the object of measurement is replaced by a measure of variable value (store of resistance, capacitance, or inductance) and, varying its value, resetting of the circuit equilibrium are added. In this case, the replacement method permits the elimination of residual non-equilibrium of the measuring circuit, the effect of magnetic and electrical fields on the circuit, mutual effects of separate elements of the circuit, and leakage and other parasitic effects.

The contraposition method consists of conducting a measurement twice, so that any cause for error in the first measurement would have opposite effect on the result of the second. The error is excluded in calculating the results of this joint measurement. For example, when weighing a mass on balance beam scales using Gauss' method, the result of the first measurement is $x = (l_2/l_1)m_1$, where l_2/l_1 is the actual ratio of the arms of the scale and m_1 is the mass of the balance weights that match the measurand. Then the object of measurement is moved to the balance pan where the weights were, and the weights are moved to where the mass was. The result of the second measurement is $x = (l_1/l_2)m_2$. Computing the square root of the product of these equalities, we derive: $x = \sqrt{m_1 m_2}$. It is evident that this method measurement makes it possible to exclude error from unequal-arm weights.

A particular case of the method of contraposition is the method of error compensation by sign. In this method, two measurements are done such that the errors enter the results with opposite sign: $x_1 = x_{\text{ctr}} + \Delta$ and $x_2 = x_{\text{ctr}} - \Delta$, where x_{ctr} is the conventional true value of the measurand, and Δ is the systematic error that must be eliminated. The error is eliminated in calculating the mean value: $\bar{x} = (x_1 + x_2)/2 = x_{\text{ctr}}$. A characteristic example of using this method is the elimination of error caused by the effect of the Earth's magnetic field. In the first measurement, the measuring instrument can be located in any position. Before the second measurement, the measuring instrument is rotated in the horizontal plane by 180°. Here the Earth's magnetic field will exert an opposite effect on the measuring instrument, and the error from magnetization is equal to the error of the first measurement but with opposite sign.

The most widespread method for eliminating systematic errors is to introduce corrections to the known components of systematic error in the measurement results. *A correction is the name given to the value of a quantity that is introduced into the unadjusted result of a measurement for the purpose of eliminating known components of systematic error (the results of measurements before corrections are called uncorrected, and afterwards are called corrected).* In accordance with the international Guide [19], the introduction of corrections for known systematic errors is an obligatory operation, preceding the processing of measurement results. Usually the algebraic addition of an unadjusted measurement result and a correction is done (taking account of its sign). In this case, the correction is equal in numerical value to the absolute systematic error and is opposite to it in sign. In those cases when the value of an absolute systematic error is proportional to the value of the measurand, $x\,(\pm\,\Delta \cong \pm\delta x)$, it is eliminated by multiplying the measurement result by the correction coefficient $K = 1 \mp \delta$. As a rule, information that is needed to determine and introduce the corrections will be apparent before the measurements. However, it is possible to determine it even after the measurement, taking account of the a posteriori (derived after the measurement) measuring information. Such an approach, in particular, is used in the run up to the acceptance of balance sheets in enterprises – the consumption of power resources, taking account of errors in the measurement facilities used for the commercial accounting.

Nevertheless, it is practically impossible to fully eliminate systematic errors. Primarily, this involves measurement methods whose systematic errors have not been studied, as well as to systematic errors that it is impossible to estimate with an actual value. This group includes, for example, measurement errors in calibrating a measuring instrument and error caused by drift of the parameters of the measuring instrument after calibration. The second group includes computational errors and errors in determining the corrections for systematic errors that have been taken into account.

Hence, after eliminating components of measurement systematic error, their residuals remain, which are called non-excluded residuals of systematic error (NRSEs) (see Par. 2.1). Not only can NRSEs not be eliminated, but they also cannot be experimentally estimated in any manner from information contained the series of measurement results, since they are present in hidden form in each result of this series. Hence one must be limited by a theoretical estimation of their limits ($\pm\,\theta$). The value of θ is usually established using an approximate calculation (for example, taking them as equal limits of allowable errors of the measuring instrument, if the random components of the measurement errors are small). If there are several reasons for the NRSEs, then they are estimated separately and then added. The method of summing the components of an NRSE are standardized in [21, 22]:

$$
\theta = \begin{cases} \sum_{i=1}^{m} |\theta_i|, & m \leq 3, \\ k\sqrt{\sum_{i=1}^{m} \theta_i^2}, & m \geq 4, \end{cases}
\tag{2.33}
$$

where $\pm \theta_i$ are the limits for the NRSE caused by the ith reason, m is the number of components of the NRSEs, k is the coefficient of dependency of the sum of component from the selected confidence probability P when they are evenly. For $P = 0.99$, $k = 1.4$, and for $P = 0.5$, $k = 1.1$.

In international practice, another method of summing is used, which shall be examined in the next chapter.

Chapter 3
Measurement Uncertainty

3.1 Error and Uncertainty

The simple and logical conception of accuracy presented in the previous chapter began to be subjected to criticism at the end of the last century in a number of foreign countries. It is the author's conviction that the fundamental reason for dissatisfaction was the term "error." The fact is that, by contrast to the Russian language, the concepts of "mistake" (i.e., a miscount or invalid action) and "error" are not differentiated in English and French ("the error" in English, "erreur" in French). For this reason, metrological terminology came into contradiction with the philosophy, accepted into general consciousness and used everywhere in the world, of the quality control of goods and services based on ISO Series 9000 standards. The essence of this methodology consists of ensuring the conditions for mistake-free execution of all productive functions and work operations. At the same time, this ideal picture is spoiled by measurement errors (in the Russian language, "pogreshnosti," having an altogether different sense), which are impossible to eliminate since they are unavoidable consequences of the limitations of measuring technology and accompany every measurement.

A similar issue arose in 1927 for the physicist Werner Heisenberg, when he was preparing to publish his noted article "On the Visualizable Content of Quantum-Theoretical Kinematics and Mechanics." In this work, he introduced into physics the well-known relationships (1) that establish the lower limits in principle of the error in measuring momentum Δp and coordinates Δx, energy ΔE, and impulse Δt:

$$\Delta p \Delta x \geq \tfrac{h}{2},$$
$$\Delta E \Delta t \geq \tfrac{h}{2}, \tag{3.1}$$

in which $h = 1.05457266 \times 10^{-34}$ is the Planck's constant. The author named these fundamental inequalities the uncertainty relationships, using the term "the uncertainty" as a synonym for the term "error."

A.E. Fridman, *The Quality of Measurements: A Metrological Reference*,
DOI 10.1007/978-1-4614-1478-0_3, © Springer Science+Business Media, LLC 2012

After the publication of this article, the term "uncertainty" began to be widely used in physics. It was used in the new concept of estimating measurement accuracy, regulated in the international document "ISO/BIPM Guide to The Expression of Uncertainty in Measurement" [19] (abbreviated "The GUM", hereinafter, the Guide). This document was published in 1993 in the name of seven authoritative international organizations:

- International Bureau of Weights and Measures (IBWM),
- International Electrotechnical Commission (IEC),
- International Federation of Clinical Chemistry (IFCC),
- International Organization for Standardization (ISO),
- International Union of Pure and Applied Chemistry (IUPAC),
- International Union of Pure and Applied Physics (IUPAP),
- International Organization of Legal Metrology (IOLM).

The Guide did in fact achieve the status of an international regulation that is obligatory for use. It is directed, first, to provide for users complete information on all components of measurement error results, and second, toward international unification of the recording of measurements and an estimate of their accuracy, with the aim of formulating a basis for international comparison of measurement results. Here it must be kept in mind that worldwide unity in the methods of estimating measurement accuracy will ensure the proper utilization of measurement results in all realms of activity.

The uncertainty concept, introduced in the Guide, consists of the following. The base concepts of the classical theory of accuracy – true value, conventional true value, and measurement error – are not introduced.[1] Instead, the concept of *measurement uncertainty* is introduced, understood as *doubt, incomplete knowledge of the value of a measurand after measurements* (interpreted in the broad sense), and as *a quantitative description of this incomplete knowledge* (interpreted in the narrow sense). Thereafter, this concept is refined: *uncertainty is a parameter associated with a measurement result and characterizing the spread of values that can be assigned to the measurand*. In mathematical statistics, two types of parameters are known that characterize the spread of non-correlated random values: the SD and the confidence interval. They are also employed as characteristics of uncertainty with the designations *standard uncertainty* and *expanded uncertainty*. Here, as one might expect, it turned out that the standard uncertainty is the complete analog of the SD of measurement errors, and expanded uncertainty is the complete analog of the confidence limits of measurement errors. And in this, the specified concept is linked with the traditional statement of the issue of estimating measurement accuracy.

Hence, in the area of practical applications, the new concept of estimating measurement accuracy has turned out to be fully identical to the classical one.

---

[1] Here it is implied that a true (conventional true) value of the quantity exists, inasmuch as it is recognized that the purpose of measurement is to find this value.

Table 1 Relationships
between quantities
characterizing measurement
error and measurement
uncertainty

X	$\lambda = L - X$	L
$\Delta_1 = I_1 - X$	$v_1 = \Delta_1 - \lambda$	$v_1 = I_1 - L$
$\Delta_i = I_i - X$	$v_i = \Delta_i - \lambda$	$v_i = I_i - L$
$\Delta_n = I_n - X$	$v_n = \Delta_n - \lambda$	$v_n = I_n - L$

In addition, these concepts are tightly associated with each other and were known in principle long ago. This becomes evident when turning to the classical work of M.F. Malikov [1]. In this monograph he presents the relationships between the quantities used in estimating measurement results. Let us introduce the following designations for their notation:

X is the unknown true value of a measurand,

I_i are the n results of single measurements of this quantity,

$L = 1/n \sum_{i=1}^{n} I_i$ is the mean of the I_i values, taken as the result of the multiple measurements of this quantity,

$\Delta_i = I_i - X$ are the n errors of the single measurements,

$v_i = I_i - L$ are the n deviations of the results of single measurements from their mean, reflecting the spread of these observations,

$\lambda = L - X$ is the random error of the result of the multiple measurements.

Table 1 presents the relationships between these quantities.

It is understood that the quantities Δ_i presented in the left section of the table are measurement errors and that the v_i presented in the right section of the table reflect the spread of observations relative to the measurement result, i.e., the uncertainty of this result. Complete symmetry is present between these quantities. Here, the values of the quantities Δ_i are completely unknown to us, while the values v_i are known from the test data. When increasing the number of observations, the mean value L converges toward the true value X. With that, their difference λ converges to zero, and the quantities v_i converge to the corresponding values Δ_i of the errors. This means that the set of values v_i is subject to the same principles as is the set of values Δ_i.

Hence, one may establish that these concepts are differentiated by the issue of which quantity pertains to the dispersion that characterizes the spread of observed values. With the classical approach, this has to do with the true value of a measurand X, but in the other case to the result L of measurement. But this distinction does not affect the behavior of the final results, since even in the classical approach the measurement errors are attributed to the measurement results. Hence both conceptions are mutually complementary, uniting into a single concept of estimation of measurement result accuracy. In this regard, following cause and effect links, it is expedient to establish the following sequence for the introduction of the basic concepts of the theory of measurement accuracy: *true value of a quantity* ⇒ *conventional true value of the quantity* ⇒ *measurement result* ⇒ *measurement error* ⇒ *indeterminacy of the measurement result, as a characteristic of this error.*

Hence the concepts of *error* and *indeterminacy* can be used harmoniously without mutual opposition.

3.2   Classification of Measurement Uncertainties

Analogous to the errors, measurement uncertainties can be classified according to several criteria.

1. Based on the method of their expression, they are divided into absolute and relative.

   Absolute measurement indeterminacy is measurement uncertainty expressed in the units of the measurand.

   Relative measurement indeterminacy of measurement results is the ration of absolute indeterminacy to the measurement results.

2. Based on the source of occurrence of measurement uncertainty, one may separate them, just as for errors, into instrumental, methodological, and subjective.
3. As follows from Sect. 2.2, based on the nature of the manifestation of error, they are separated into systematic, random, and gross errors. The Guide has no classification for uncertainties using this criterion. It is stated, at the very beginning of this document, that before statistical processing of a series of measurements, all known systematic errors must be excluded from them. Hence the separation of uncertainties into systematic and random is not introduced. In its place is separation of uncertainties into two types, based on the method of evaluation:

   - *uncertainty from type A evaluation (type A uncertainty) is uncertainty that is evaluated by statistical methods,*
   - *uncertainty from type B evaluation (type B uncertainty) is uncertainty that is evaluated by non-statistical methods.*

   Correspondingly, two methods of evaluation are presented:

   - *type A evaluation is the derivation of statistical estimates based on the results of a series of measurements,*
   - *type B evaluation is the derivation of estimates based on a priori non-statistical information.*

   At first glance, it seems that this innovation merely consists of replacing the existing terms of known concepts with others. Actually, one can evaluate only the random error with statistical methods, and hence type A uncertainty is what was earlier called random error. Analogously, non-statistical random error can be estimated only on the basis of a priori information, and hence there is likewise a mutually unique correspondence between type B uncertainty and non-statistical random error.

   However, from the author's point of view, the introduction of these concepts is altogether reasonable. The fact is that in measurements using complex methods, including a great number of sequentially executed operations, it is essential to evaluate and account for the great number of sources of the uncertainty of the final result. Here, their division into non-statistical random error and random can be falsely oriented. We shall introduce two examples.

Table 2 Terms and approximate analogs of the concept of uncertainty with the classical theory of accuracy

Classical theory	Concept of uncertainty
Error of a measurement result	Uncertainty of a measurement result
Random error	Uncertainty evaluated using type A
Non-statistical random error	Uncertainty evaluated using type B
SD (standard deviation) of the error of a measurement result	Standard uncertainty of a measurement result
Confidence limits of a measurement result	Extended uncertainty of a measurement result
Confidence probability	Coverage probability
Quantile (coefficient) of error distribution	Coverage factor

Example 3.1. A substantial part of the uncertainty of an analytical measurement can consist of the definition of the calibration function of an instrument, which is non-statistical random error at the time of the measurement. Consequently, it is necessary to evaluate it based on a priori information using non-statistical methods. However, in many analytical measurements the basic source of this uncertainty is random error in the weighing process in preparing the calibration mixture. To improve measurement accuracy, one may apply multiple weighings of this standard sample and find an estimate of the error of this weighing process using statistical methods. This example shows that in some measurement techniques, for the purpose of improving the accuracy of measurement results, a number of systematic components of measurement uncertainty can be evaluated by statistical methods, i.e., they are type A uncertainties.

Example 3.2. For a number of reasons, such as to save on production expenses, the measurement method may provide for conducting no more than three single measurements of each quantity. In this case, the measurement result can be defined as the mean, mode, or median of the values obtained, but statistical methods of evaluating uncertainty with this sample size will provide only a very coarse estimate. It is more reasonable to provide an a priori calculation using standardized indexes of the accuracy of the measurement instrument; i.e., this is a type B evaluation. Consequently, in this example, by contrast with the preceding one, the uncertainty of the result of measurements, a significant part of which is caused by the effect of factors of a random nature, will be type B uncertainty.

Along with this, the traditional division of errors into systematic, non-statistical random errors, and random also loses meaning since it more precisely reflects other criteria: the nature of its appearance in the measurement result and the causative link with effects that are sources of error. Hence, the classification of uncertainties and errors of measurements are not alternatives, but complement each other.

The Guide also has several other technical innovations. A summary table of terminological distinctions between the concept of uncertainty and classical theory of accuracy is presented below (Table 2).

New terms shown in this table have the following definitions.

1. *Standard uncertainty is uncertainty expressed in the form of a standard deviation;*
2. *Extended uncertainty is a value specifying the interval about the measurement result, within which limits, as expected, are found the greatest part of the values of the distribution that can be ascribed to the measurand with sufficient cause.*

   1. Notes.
      With each value of extended uncertainty is associated the value of its coverage probability P.
   2. An analog of extended uncertainty is the confidence limits of measurement error.

3. *Coverage probability is the probability that, in the opinion of the experimenter, corresponds with the extended uncertainty of the result of measurements.*

   1. Notes.
      An analog of this term is confidence probability, corresponding to the confidence limits of an error.
   2. Coverage probability is selected taking account of information on the type of distribution law for the uncertainty.

4. *A coverage factor is a coefficient that depends on the type of distribution of uncertainty of the measurement result and of the coverage probability, and is numerically equal to the ratio of the extended uncertainty corresponding to the specified probability of coverage, to the standard uncertainty.*
5. *The number of degrees of freedom is a parameter of the statistical distribution, equal to the number of independent connections to the statistical sample being evaluated.*

3.3   Procedure for Evaluation of the Uncertainty of the Measurement Result

Insofar as it pertains to the terminological innovations of the Guide, one must recognize that its introduction had a huge positive effect and made it possible to increase confidence in the evaluations of the accuracy of measurement results, as used in all countries of the world in product certification and in settling mutual accounts between suppliers and purchasers; and in public health, scientific research, and other areas of activity. This effect is due to the regulation of a unified method of evaluating measurement results and their uncertainties, and their mass utilization by measuring and testing laboratories. Secondly, this effect was caused by the fact that the methodology regulated by the Guide contains very high requirements for evaluating accuracy than had been practiced earlier. In this regard, let us examine in detail this methodology in its current exposition [23].

Evaluating the result of measurements and its uncertainty takes place as follows:

- a measurement equation is formulated,
- the input values and their standard uncertainties are evaluated,
- the output values and their standard uncertainties are evaluated,
- the uncertainty budget is formulated,
- the extended uncertainty of a measurement result is evaluated, and the measurement result is presented.

Let us examine these valuation steps.

1. Formulation of a measurement equation.
   A measurement equation is understood to be a mathematical relationship between the measurands $X_1, \ldots, X_k$ and the measurement result Y

$$Y = f(X_1, \ldots, X_k, \ X_{k+1}, \ldots, X_n), \tag{3.2}$$

where $X_1, \ldots, X_n$ are quantities influencing Y. This pertains both to the directly measurable quantities $X_1, \ldots, X_k$, and to all other influencing values $X_{k+1}, \ldots, X_n-$ reference data, constants, adjustments, etc. In the methodology being studied, these are called input quantities (*an input quantity of a measurement equation or of an equation used to evaluate the uncertainty of this measurement*). In contrast to these, a measurement result is called an output quantity (*an output quantity of a measurement equation is a quantity, the value of which is determined in measurement*).

   An equation of measurements is formulated as follows. The functional dependence of the measurement result Y on the measurand is written as

$$Y = f(X_1, \ldots, X_k), \tag{3.3}$$

which is the mathematical description of a physical effect at the foundation of the measurement methods.

   Then the measurement conditions and other factors that influence the measurement results are analyzed. The quantities that describe these factors are included in (3.3). One must keep in mind that the factors influencing the measurement result and its accuracy must be fully accounted for, as much as possible, in the measurement equation. Hence if, for example, an adjustment is introduced into the measurement result, the value of which is equal to zero in a specific case, then this adjustment must nevertheless figure in the measurement equation as an input quantity, since the uncertainty of its quantity contributes to the total uncertainty of the measurement.

2. Evaluation of the input quantities and their standard uncertainties.
   The mean estimate of the supposed distribution of its values is taken as the value of the input quantity, and the estimation of the standard deviation of this mean estimate is taken as its standard uncertainty.

   If there are results $x_{i1}, \ldots, x_{im}$ of the independent measurements of one of the input quantities $X_i, \ i = 1, \ldots, n$, conducted under the same conditions, the evaluation of type A uncertainty, i.e., statistical estimation, is employed. As follows from the preceding chapter, with normal distribution of the measurement results, the best estimate x_i of this quantity is the mean

$$x_i = \bar{x}_i = \frac{1}{m} \sum_{j=1}^{m} x_{ij}, \tag{3.4}$$

and the standard uncertainty of this estimate is equal to the SD of the mean

$$u_A(x_i) = u_A(\bar{x}_i) = \sqrt{\frac{1}{m(m-1)} \sum_{j=1}^{m} (x_{ij} - \bar{x}_i)^2}. \tag{3.5}$$

Estimates of these quantities with the generalized normal distribution of the measurement results are presented in 4.2.

The following sources of a priori information are the initial data for type B evaluation of the quantity and its standard uncertainty:

– data from previous measurements of this quantity contained in measurement protocols, evidence from calibration and testing, or other documents;
– standards for measurement accuracy as specified in technical documentation on measurement methods and the measurement instrument;
– values of constants and reference data, and their uncertainties;
– information on the supposed distribution of the values of the quantity, available in technical reports and bibliographical sources;
– the researcher's experience or knowledge regarding general principles to which the properties of applied materials or instruments are subject.

The following cases of type B evaluation are differentiated.

(a) If only one value X_i is known, such as the result of a single measurement, adjustment, or reference data, then this value is taken as an estimate x_i. An evaluation of the standard uncertainty $u_B(x_i)$ is found as follows:

– if an estimate of the standard uncertainty $u(x_i)$ is known, then $u_B(x_i) = u(x_i)$;
– if the extended uncertainty $U(x_i)$ and coverage factor k are known, the standard uncertainty is calculated from the formula

$$u_B(x_i) = U(x_i)/k. \tag{3.6}$$

If the coverage factor is not specified, it will be accepted by taking account of the current hypothesis regarding the form of the distribution of the uncertainty of the quantity X_i and the coverage probability, to which the extended uncertainty $U(x_i)$ corresponds. Table 3 shows some typical cases.

(b) If there are assumptions regarding the probability distributions of the quantity X_i, then the mean and standard deviation of this distribution are taken as the estimate of x_i and its standard uncertainty $u_B(x_i)$.

Table 3 Coverage factors of distributions of uncertainty of input quantities

Surmised distribution of the uncertainty of the input quantity	Coverage probability P to which $U(x_i)$ corresponds	Coverage factor k
Equal-probability distribution	0.99–1.0	1.73
"	0.95	1.65
Normal distribution	1 (limit of allowable values)	3.00
"	0.99 ($U(x_i)$ of primary and secondary standards)	2.60
"	0.95 ($U(x_i)$ of operational standards)	2.00
Distribution unknown	–	2.00

(c) If only the upper a_+ and lower a_- bounds for the values of the quantity X_i can be estimated, then the equal-probability distribution is accepted. In this case

$$x_i = \frac{1}{2}(a_+ + a_-), \tag{3.7}$$

$$u_B(x_i) = \frac{a_+ - a_-}{2\sqrt{3}} \tag{3.8}$$

If $a_+ = -a_- = a$, then $x_i = a$, $u_B(x_i) = a/\sqrt{3}$.

3. Evaluation of output quantities and their uncertainties

The result y of measurements is calculated by formula (3.2), in which are substituted the values x_i of input values as defined in the preceding stage:

$$y = f(x_1, ..., x_k, x_{k+1}, ..., x_n). \tag{3.9}$$

Expanding the function $y = f(x_1, ..., x_n)$ into a Taylor series in the neighborhood of the point $(x_1, ..., x_n)$, we derive the dependence of the error Δy of the result y of measurements on the errors Δx_i of the estimates x_i of input quantities:

$$\Delta y = \sum_{i=1}^{n} \frac{\partial f(x_1, ..., x_n)}{\partial x_i} \Delta x_i + o(\Delta x),$$

where $\frac{\partial f(x_1, ..., x_n)}{\partial x_i} = \frac{\partial f(X_1, ..., X_n)}{\partial X_i} |X_s = x_s$, $s = 1, ..., n$ is the partial differential of the function $f(X_1, ..., X_n)$ for quantity X_i, and is calculated for expected values of all input quantities.

Consequently, the contribution $u_i(y)$ of each input quantity X_i to the uncertainty y of the measurement result is defined by the formula

$$u_i(y) = c_i u(x_i), \tag{3.10}$$

where $c_i = \frac{\partial f(x_i, ..., x_n)}{\partial x_i}$ is the coefficient of sensitivity of the output quantity Y to the input quantity X_i.

Coefficients of sensitivity can be defined in various manners: directly by formula (3.10), by numerical differentiation of the measurement equation, and by experimental study.

If the estimates of the input quantities are correlated, then the coefficients of their mutual effect $c_{ij} = \frac{\partial^2 f(x_1,...,x_n)}{\partial x_i \partial x_j}$ and correlation coefficients

$$r(x_i, \ x_j) = \frac{u(x_i, x_j)}{u(x_i) u(x_j)},\tag{3.11}$$

are found, where $u(x_i, \ x_j)$ is the covariance of quantities x_i and x_j.

The covariance $u(x_i, \ x_j)$ can be estimated if there are m pairs of results $x_{is}, \ x_{js} \ (s = 1, ..., n)$ of independent repeated measurements of the two quantities x_i and x_j. In this case, this is equal to

$$u(x_i, \ x_j) = \frac{1}{m(m-1)} \sum_{s=1}^{m} (x_{is} - \bar{x}_i)(x_{js} - \bar{x}_j).\tag{3.12}$$

Now all is in readiness to calculate the total standard uncertainty $u(y)$ of the result of measurements. In the general case, with correlated estimates of input values, this is calculated using the formula

$$u(y) = \sqrt{\sum_{i=1}^{n} u_i^2(y) + \sum_{i,j=1, \ i\neq j}^{n} c_{ij} r(x_i, \ x_j) u_i(y) u_j(y)},\tag{3.13}$$

With uncorrelated estimates of input values $r(x_i, \ x_j) = 0$, and consequently,

$$u(y) = \sqrt{\sum_{i=1}^{n} u_i^2(y)}.\tag{3.14}$$

It is permissible to consider the correlation of two input values as negligibly small if:

– these values are mutually independent (for example if they are observed many times but not at the same time, and in different tests independently of each other);
– one of these values can be considered to be a constant;
– there are no reasons for correlation between these values.

Nevertheless, ignoring correlation between input values may lead to a mistaken estimation of the standard uncertainty of a measurand. Hence, if the degree of correlation between values X_i and X_j is unknown, it is useful to estimate the maximum effect of this correlation on a measurement result, using an estimate of the maximum standard uncertainty of the measurand. This estimate is equal to

$$u^2(y) \leq (|u_i(y)| + |u_j(y)|)^2 + u_r^2(y),$$ (3.15)

where $u_r(y)$ is the contribution to standard uncertainty of the measurand from the remaining input values that are considered uncorrelated.

1. Notes.
   If $f(X_1, ..., X_n) = \sum_{i=1}^{n} \varphi_i(X_i)$, then the estimate of the output value is equal to $y = \sum_{i=1}^{n} \varphi_i(x_i)$, and its absolute standard uncertainty is $u(y) = \sqrt{\sum_{i=1}^{n} (\partial \varphi_i(x_i)/\partial x_i)^2 u^2(x_i)}$. In the special case for $\varphi_i(X_i) = p_i X_i$, $i = 1, ..., n$, we derive $u(y) = \sqrt{\sum_{i=1}^{n} p_i^2 u^2(x_i)}$.
2. If $f(X_1, ..., X_n) = \prod_{i=1}^{n} \varphi_i(X_i)$, then the estimate of the output value is equal to $y = \prod_{i=1}^{n} \varphi_i(x_i)$, and its relative standard uncertainty is

$$u_{\text{rel}}(y) = \frac{u(y)}{y} = \sqrt{\sum_{i=1}^{n} \left(\frac{\partial \varphi_i(x_i)/\partial x_i}{\varphi_i(x_i)} \right)^2 u^2(x_i)}.$$

In the special case for $\varphi_i(X_i) = X_i^{p_i}$, $i = 1, ..., n$ we derive $u_{\text{rel}}(y) = \sqrt{\sum_{i=1}^{n} p_i^2 u_{\text{rel}}^2(x_i)}$, where $u_{\text{rel}}(x_i) = u(x_i)/x_i$ is the relative standard uncertainty of the value X_i.

4. Formulation of the uncertainty budget.
   The term "uncertainty budget" is understood to mean a formalized brief exposition of the evaluation of a measurement's uncertainty. (*"Uncertainty budget" is the designation given to a complete list of sources of measurement uncertainty, specifying its standard uncertainty and contribution to the total standard uncertainty of the result of measurements.*).
   An uncertainty budget is presented in the form of table (see Table 4).
   Column 1 in this table enumerates the input quantities of the measurement equation, and column 2 the estimates of input quantities as derived either as a result of measurements or based on other information. Column 3 presents the values of the standard uncertainties of these estimates. Column 4 indicates the type of evaluation of the uncertainty. When needed, it also shows the suggested distribution law of the estimates. For example, if the estimate of the value is derived from results of many measurements, then as a rule the normal distribution law of its values and type A evaluation are proposed. Column 5 presents the coefficients of sensitivity of input quantities $c_i = \partial f/\partial x_i$, and column 6 shows the values of the quantities $u_i(y) = |\partial f/\partial x_i| \times u(x_i)$ for the input values to the total standard uncertainty $u(y)$ (the product of the values from column 3 and the absolute magnitude of the value from column 5).

Table 4 Recommended form for presenting an uncertainty budget

1	2	3	4	5	6
Quantity	Estimate	Standard uncertainty	Type of estimate (distribution law)	Coefficient of sensitivity	Contribution to total standard uncertainty
X_1	x_1	$u(x_1)$	A (B)	$c_1 = \frac{\partial f}{\partial x_1}$	$u_1(y) = \left\lvert \frac{\partial f}{\partial x_1} \right\rvert \times u(x_1)$
X_n	x_n	$u(x_n)$	A (B)	$c_n = \frac{\partial f}{\partial x_n}$	$u_n(y) = \left\lvert \frac{\partial f}{\partial x_n} \right\rvert \times u(x_n)$
Y	$y = f(x_i)$	$u(y)$			$u(y) = \sqrt{\sum_{i=1}^{n} u_i^2(y)}$

The last row of the table shows the result of measurements y and its standard uncertainty $u(y)$, equal to the mean square sum of all values presented in column 6. All values of quantities that are presented in the table include the designations of the units of these quantities.

5. Defining extended uncertainty. Presenting the result of a measurement

The extended uncertainty $U(y)$ is equal to the product of the standard uncertainty $u(y)$ of the measurement of an output quantity y with the coverage factor k:

$$U(y) = ku(y). \tag{3.16}$$

The Guide recommends taking the coverage factor to be equal to the quantile of Student's distribution with a coverage probability of $P = 0.95$ and number of effective degrees of freedom ν_{eff}:

$$k = t_{P=0.95}(\nu_{\text{eff}}), \tag{3.17}$$

and it is proposed to determine the number of effective degrees of freedom using the Welch-Satterwaite formula

$$\nu_{\text{eff}} = \frac{u^4(y)}{\sum_{i=1}^{n} u_i^4(y)/\nu_i}, \tag{3.18}$$

in which ν_i is the number of degrees of freedom of the estimate of the contribution of the i-th input quantity to the uncertainty of the result of measurements.

Hence the coverage factor is selected from tables of Student's distribution for a probability $P = 0.95$ and number of degrees of freedom equal to ν_{eff}, rounded down to the highest integer less than ν_{eff}. These values are shown in Table 5.

In estimating the contribution of the type A uncertainty of quantity X_i to the uncertainty of the measurement results, the number of degrees of freedom will be taken as equal to $\nu_i = m_i - 1$, where m_i is the number of repeated measurements of this quantity.

Table 5 Coverage factors k for various degrees of freedom v_{eff}

v_{eff}	1	2	3	4	5	6	7	8	10	20	50	∞
k	13.97	4.53	3.31	2.87	2.65	2.52	2.43	2.37	2.28	2.13	2.05	2.00

In estimating the contribution of the type B uncertainty, which is subject to uniform distribution, the number of degrees of freedom will be taken as equal to infinity: $v_i = \infty$.

In practice, type A evaluation of the uncertainty is often just for one input quantity, and the uncertainties of all other input quantities are type-B evaluated in accordance with their uniform distribution. It is easy to show that in this case, the formula (3.18) is simplified:

$$v_{\text{eff}} = (m-1) \times \frac{u^4(y)}{u_A^4(y)}, \tag{3.19}$$

where $u_A(y)$ is the contribution to the total standard uncertainty of the quantity for which there is type A estimation of uncertainty, and m is the number of repeated measurements of this quantity.

If there is a basis for assuming a normal distribution for the probabilities of the measurand y, the coverage factor will be taken as equal to $k = 2$. Here, the extended uncertainty of the measurement result approximately corresponds to the covering probability of 0.95.

In those cases when information is lacking on the form of the distribution of the uncertainty for the measurand, the Guide recommends for the sake of international unification that one likewise take this as $k = 2$, and consider that here the extended uncertainty of the measurement result will approximately correspond to the coverage probability of 0.95.

In the records and certificate of measurement, one must indicate the measurement result y, the extended uncertainty $U(y)$ associated with it, and the coverage factor k. For example: "The measurement result of the length of the item is 153.2 mm. The extended uncertainty of the measurement result is ± 1.4 mm with a coverage factor equal to 2".

Another formulation is also possible, in which one shows the interval $(y - U(y), y + U(y))$ in which the value of the measurand is found, with the corresponding coverage factor. For example: "Measurements have shown that the length of the item is in the interval (151.8–154.6) mm with a coverage factor equal to 2".

Example 3.3. Measuring net weight of oil by indirect method

The net weight of oil is measured indirectly in accordance with the measurement equation

$$M_n = V(1 - \varphi_w)g[1 - (x_{cs} + x_{\text{sol}})]T,$$

Table 6 Uncertainty budget for measurements of net weight of oil

1	2	3	4	5	6
		Relative standard uncertainty	Type of evaluation (distribution law)	Coefficient of sensitivity	Contribution to total standard uncertainty
Quantity	Estimate				
V	1,200.0 m³	0.001	B (normal law)	1	0.0010
φ_w	0.011	0.0067	Same	0.011	0.000074
g	0.81 t/m³	0.0033	–	1	0.00330
x_{cs}	0.002	0.01	–	0.002	0.000020
x_{sol}	0.01	0.0067	–	0.0105	0.000070
M_n	949.8 t				$u_{rel}(M_n) = 0.00345$

in which V is the volume of fluid in m³, φ_w is the volume percent of water, g is the density of dewatered oil in t/m³, and x_{cs} and x_{sol} are the mass shares of chloride salts and solids.

Measurement results are: $V = 1200.0\,\text{m}^3$, $\varphi_w = 0.011$, $g = 0.81\,\text{t/m}^3$, $x_{cs} = 0.002$, $x_{sol} = 0.01$.

The measurement accuracy of these quantities is characterized by the normal error distribution and limits of allowable relative error: $\delta V = 0.003$, $\delta\varphi_w = 0.02$, $\delta g = 0.01$, $\delta x_{cs} = 0.03$, $\delta x_{sol} = 0.02$.

Measurement result: $M_n = 1200(1 - 0.011)0.81[1 - (0.002 + 0.01)] = 949.8\,t$.

The uncertainty of the measurements of all input quantities is characterized by the limits of allowable relative error. Consequently, we evaluate them as type B and, taking account of the assumption of the normal law, we obtain coverage factor $k = 3$ in accordance with Table 3. Hence the standard uncertainties of the input quantities are:

$$u_{1.rel} = u_{rel}(V) = 0.001, \quad u_{2.rel} = u_{rel}(\varphi_B) = 0.0067 \quad u_{3.rel} = u_{rel}(g) = 0.0033,$$
$$u_{4.rel} = u_{rel}(x_{x.c}) = 0.01, \quad u_{5.rel} = u_{rel}(x_{m.n}) = 0.0067.$$

Let us find the coefficients of sensitivity of the measurement results to a change in the input quantities. In accordance with Note 2 of section 3, the relative standard uncertainty of the measurement results is calculated using the formula

$$u_{rel}(M_{\mathcal{H}}) = \sqrt{\sum_{i=1}^{5} P_i^2 u_{i.rel}^2},$$

where $p_1 = p_3 = 1$, $p_2 = \frac{\varphi_w}{1-\varphi_w} = 0.011$, $p_4 = \frac{x_{cs}}{1-x_{cs}-x_{sol}} = 0.002$, $p_5 = x_{sol}/1 - x_{cs} - x_{sol} = 0.0105$.

Table 6 shows the uncertainty budget of this measurement.

The extended uncertainty of the measurement results is evaluated by starting with the normal distribution (since all components were distributed according to this law) and coverage probability $P = 0.95$. In this case $k = 2$ and the relative extended uncertainty $U_{rel}(M_n) = 2 \cdot 0.00345 = 0.0069 \cong 0.7\%$.

Hence, the result of measuring the net weight of the oil is 949.8 t. The relative extended uncertainty of this result is 0.7% with a coverage factor of 2, corresponding to a coverage probability of $P = 0.95$.

Example 3.4. Calibrating a resistor with a nominal value of 10 kΩ [23]

The electrical resistance of a four-terminal resistor is measured with a numerical 7.5-digit voltmeter while also measuring the electrical resistance of a reference four-terminal resistor that has the same nominal electrical resistance. The resistors are immersed in an oil bath where the temperature is monitored by a glass mercury thermometer. The nominal temperature of the oil bath is 23°C. Measurements are done by the substitution method: the four-terminal connectors of each resistor are connected in order to the terminals of the voltmeter. It has been established that a current of 100 mA cannot cause any noticeable heating of resistors that have a nominal value of 10 kΩ. The effect of external leakage on the measurement results of electrical resistances can likewise be considered insignificant.

1. Measurement equation:

$$R_X = (R_{nE} + \delta R_{\partial E} + \delta R_{TE})r_c r - \delta R_{TX},$$

where R_X is the electrical resistance of the resistor being calibrated at the nominal temperature,

R_{nE} is the value of electrical resistance assigned to the reference resistor at its last calibration,

$\delta R_{\partial E}$ is the drift of the electrical resistance of the reference resistor after its last calibration,

$\delta R_{TE}, \delta R_{TX}$ is the change to the reference resistor and the resistor being calibrated, due to the temperature difference from the nominal temperature,

$r = \hat{R}_X/\hat{R}_E$ is the ratio of the indicated values of the electrical resistances of the resistor being calibrated and the reference resistor ($\hat{R}_X = R_X + \delta R_{TX}, \hat{R}_\ni = R_{nE} + \delta R_{\delta E} + \delta R_{TE}$),

r_c is the adjustment for stray voltages and the resolution capability of thevoltmeter.

2. Input quantities and their standard uncertainties

(a) Reference: In the certificate of calibration of the reference resistance, the value $R_S = 10,000.053\Omega \pm 5$ mΩ (coverage factor $k = 2$) is presented, at a reference temperature of 23°C. Hence the standard uncertainty of the assigned reference value is $u_1 = u_B(R_E) = 5/2 = 2.5$ mΩ.

(b) Drift of the value of a reference resistor after its last calibration: expected value $\delta R_D = \pm 20$ mΩ. Uncertainty of this estimate is type-B evaluated and described by a uniform distribution in the interval ± 10 mΩ. Its standard uncertainty is calculated by formula (3.8): $u_2 = u_B(\delta R_{\partial E}) = 10/\sqrt{3} = 5.8$ mΩ.

(c) The temperature of the oil bath is monitored with a calibrated thermometer. Taking account of the metrological characteristics of the thermometer and the temperature gradient in the oil bath, it is considered that the temperature deviations of the resistors from the specified temperature of 23°C do not exceed ±0.055 K. The value of the temperature coefficient (TC) of the reference resistor, $5 \times 10^{-6} \mathrm{K}^{-1}$, provides a deviation of electrical resistance as no greater than $\delta R_{TE} = \pm 10^7 \times 5 \times 10^{-6} \times 0.055 = \pm 2.75 \, \mathrm{m\Omega}$. In accordance with the technical documentation on the resistor being calibrated, its TC is no greater than $10 \times 10^{-6} \, K^{-1}$. Consequently, the deviation of the resistor being calibrated due to temperature variations must not exceed $\delta R_{TX} = \pm 5.5 \, \mathrm{m\Omega}$. Their standard uncertainties are likewise type-B evaluated, and described by a uniform distribution with coverage probability 0.99-1.0 (since the limits of possible deviations are indicated): $u_3 = u_B(\delta R_{TE}) = 2.75/\sqrt{3} = 1.6 \, \mathrm{m\Omega}$, $u_4 = u_A(\delta R_{TX}) = 5.5/\sqrt{3} = 3.2 \, \mathrm{m\Omega}$.

(d) Since the same voltmeter is used to measure R_{iX} and $R_{iE,}$ their contributions to the uncertainty are correlated. Measuring the ratio of resistances reduces the correlation, since only the relative difference between readings is brought to attention. In this respect the action of systematic effects (stray voltages and instrument resolution), which are estimated within the limits $\pm 0.5 \times 10^{-6}$ for each reading, is significantly reduced. The resultant distribution of the ratio r_c is equal-probability with mean 1.0000000 and limits $1.0000000 \pm 1.0 \times 10^{-6}$. Hence the standard uncertainty caused by these factors is equal to

$$u_5 = u_B(r_c) = \frac{1 \times 10^{-6}}{1.73} = 0.58 \times 10^{-6}.$$

(e) Five measurements of the ratio r were taken: 1.0000104; 1.0000107; 1.0000106; 1.0000103; and 1.0000105. The measurement result is
$\bar{r} = 1.0000100 + \frac{4+7+6+3+5}{5} \times 10^{-7} = 1.0000105$, and its standard uncertainty

is $u_6 = u_A(r) = \sqrt{\frac{1}{4 \cdot 5}(1^2 + 2^2 + 1^2 + 2^2 + 0)} \times 10^{-7} = 0.071 \times 10^{-6}$.

(f) Coefficients of sensitivity

For the values r_c and r the coefficient of sensitivity $c_i = \partial R_X/\partial r \cong \hat{R}_E r_c \cong 10,000 \, \Omega$. For other values, $c_i = r_c r \cong 1.0$.

(g) The uncertainty budget is shown in Table 7.

(h) Extended uncertainty:

$$U(R_X) = ku(R_X) = 2 \times 9.28 \cong 19 \, \mathrm{m\Omega}.$$

(i) Final results: the result of measurements of the electrical resistance of a resistor of nominal value 10 kΩ at 23°C is $(10,000.178 \pm 0.019) \, \Omega$. The claimed extended uncertainty of the measurement is established as the

Table 7 Uncertainty budget in calibrating a resistor

Parameter X_i	Estimate x_i	Standard uncertainty $u_i = u(x_i)$	Probability distribution	Coefficient of sensitivity c_i	Contribution $u_i(R_X)$
R_{nE}	10,000.053 Ω	2.50 mΩ	Equal probability	1	2.50 mΩ
$\delta R_{\partial E}$	0.020 Ω	5.80 mΩ	Same	1	5.80 mΩ
δR_{TE}	0.000 Ω	1.60 mΩ	"	1	1.60 mΩ
δR_{TX}	0.000 Ω	3.20 mΩ	"	1	3.20 mΩ
r_c	1.0000000	0.58×10^{-6}	"	10,000 Ω	5.80 mΩ
r	1.0000105	0.07×10^{-6}	normal	10,000 Ω	0.71 mΩ
R_X	10,000.178 Ω			$u(R_X) =$	9.28 mΩ

standard uncertainty of the measurement, multiplied by the coverage factor $k = 2$, which for a normal distribution corresponds approximately to the coverage probability 0.95.

Chapter 4
Methods for Estimation of Measurement Results and Their Uncertainties

4.1 Elimination of Outliers from the Measurement Series

A set of measurement results x_1, ..., x_n can have a value that differs strongly from others. For example, there is some minimum value x_{min} that is significantly less than the others, or some maximum value x_{max} that is significantly greater than the others. These values might be the result of a low-probability event – a random error subject to the random error distribution of this measurement process, but significantly differing in magnitude from the other members of the measurement series. But they also might be the result of mistakes, as mentioned in Sect. 2.1. In this regard, the need arises to discover and eliminate the measurement results that are incompatible in this sense and could distort the statistical conclusions.

Such measurement results are called sharply divergent results or outliers. In principle, an outlier in a set of number is taken to be a number that in some well-defined sense is incompatible with the others [24]. In processing the results of measurements, this concept is defined more concretely: *the word "outlier" is understood to mean a result for which the statistic pertaining to the selected compatibility criterion exceeds a critical value that corresponds to the 99% confidence interval.* The concept of a *quasi-outlier* is also introduced. *This term is understood to mean a result for which the statistic pertaining to the selected compatibility criterion exceeds a critical value that corresponds to the 95% confidence interval, but does not exceed a critical value that corresponds to the 99% confidence interval.*

In mathematical statistics, a great number of compatibility criteria have been developed. Let us examine the Grubbs test, which is one of the most widely used criteria in measurement practice. In accordance with this criterion, the measurement results subject to analysis are first arranged in increasing order, forming a monotonic series $\{x_i\}$, $i = 1$, ..., n, in which $x_1 = x_{min}$ and $x_n = x_{max}$ are the extreme values. The Grubbs statistic has the form:

$$G_n = \frac{x_n - \bar{x}}{S}, \quad G_1 = \frac{\bar{x} - x_1}{S}, \tag{4.1}$$

A.E. Fridman, *The Quality of Measurements: A Metrological Reference*,
DOI 10.1007/978-1-4614-1478-0_4, © Springer Science+Business Media, LLC 2012

Table 8 Critical values of the Grubbs statistics [25]

n	Single largest or single smallest		Two largest or two smallest	
	Above 1%	Above 5%	Above 1%	Above 5%
3	1.155	1.155	–	–
4	1.496	1.481	0.000	0.000
5	1.764	1.715	0.001	0.009
6	1.973	1.887	0.011	0.034
7	2.139	2.020	0.030	0.070
8	2.274	2.126	0.056	0.110
9	2.387	2.215	0.085	0.149
10	2.482	2.290	0.115	0.186
11	2.564	2.355	0.144	0.221
12	2.636	2.412	0.173	0.253
14	2.755	2.507	0.228	0.311
16	2.852	2.585	0.276	0.360
18	2.932	2.651	0.320	0.402
20	3,001	2.709	0.358	0.439
25	3.135	2.822	0.437	0.512
30	3.236	2.908	0.498	0.567
40	3.381	3.036	0.586	0.644

where

$$\bar{x} = \frac{1}{n}\sum_{i=1}^{n} x_i, \quad S = \sqrt{\frac{1}{n-1}\sum_{i=1}^{n}(x_i - \bar{x})^2}$$

are the mean and standard deviation of the series, respectively. Table 8 presents the threshold values of the Grubbs statistic. An outlier or quasi-outlier is established when the values of the Grubbs statistics exceed the values of the 1 and 5% critical values presented in this table, respectively.

If it turns out in this test that a series of measurements does not have any outliers, it is advisable to additionally check the two maximum and two minimum results for outliers. The Grubbs statistic for the joint check of the two maximum results has the form

$$G = \frac{S_{n-1,n}^2}{S_0^2}, \tag{4.2}$$

where $S_0^2 = \sum_{i=1}^{n}(x_i - \bar{x})^2$, $S_{n-1,n}^2 = \sum_{i=1}^{n-2}(x_i - \bar{x}_{n-1,n})^2$, $\bar{x}_{n-1,n} = \frac{1}{n-2}\sum_{i=1}^{n-2} x_i$.

Analogously, the Grubbs statistic for the joint check of the two minimum results has the form

$$G = \frac{S_{1,2}^2}{S_0^2}, \tag{4.3}$$

Table 9 Recommended number of intervals for grouping

Count of experimental data	Recommended number of intervals
25–100	6–9
100–500	8–12
500–1,000	10–16
1,000–10,000	12–22

where $S_{1,2}^2 = \sum_{i=3}^{n} (x_i - \bar{x}_{1,2})^2$, $\bar{x}_{1,2} = \frac{1}{n-2} \sum_{i=3}^{n} x_i$.

An outlier or quasi-outlier is established when the values of the Grubbs statistic are less than the 1 or 5% critical values shown in Table 8, respectively.

Example 4.1. The results of measurements of the mass fraction of sulfur in coal, in percentages, are shown as the following monotonic series: $x_1 = 0.560$; $x_2 = 0.567$; $x_3 = 0.577$; $x_4 = 0.580$; $x_5 = 0.590$; $x_6 = 0.703$; $x_7 = 0.708$; $x_8 = 0.733$. The mean and standard deviation of the series are:

$$\bar{x} = \frac{1}{8} \sum_{i=1}^{8} x_i = 0.690\ \%, \quad S = \sqrt{\frac{1}{7} \sum_{i=1}^{8} (x_i - 0.690)^2} = 0.024\ \%.$$

The Grubbs statistics are equal to

$$G_8 = \frac{0.733 - 0.690}{0.024} = 1.792, \quad G_1 = \frac{0.690 - 0.660}{0.024} = 1.25.$$

Since they are less than the 1 and 5% critical values for $n = 8$ (2.274 and 2.126, respectively), the analyzed sample does not contain outliers and quasi-outliers. We do not conduct the check for outliers of the two largest and two smallest results because of the small sample size.

4.2 Check of the Hypothesis about the Normal Distribution of Experimental Data

It is possible to verify the hypothesis that the distribution of the experimental data does not contradict the theoretical distribution, using a variety of criteria. Below we shall examine the most well-known of them – the Pearson "chi-square" test. This test provides good results for experimental data numbering $n \geqslant 50$, although it is often used even for smaller samples sizes. It is good practice to calculate it using the following procedure.

1. The data are grouped into intervals. Table 9 shows the recommended number of intervals, l.

In addition, one must keep in mind that the length of a grouping interval,

$$h = \frac{x_{max} - x_{min}}{l},$$ (4.4)

where x_{max}, x_{min} are the largest and smallest values in the experimental data set, must be greater than the rounding error in recording this data. The value h calculated by this formula is rounded. Then the interval midpoints $\hat{x}_i$ and the corresponding empirical frequencies (number of instances in the interval) $\hat{n}_i$.

2. The mean $\bar{x}$ and SD S are calculated.
3. For each interval, define

$$z_i = \frac{\hat{x}_i - \bar{x}}{S}$$ (4.5)

and the theoretical number of instances of experimental data in the interval

$$n_i = n\frac{h}{S}f(z_i),$$ (4.6)

where $f(z_i) = 1/\sqrt{2\pi}e^{-0.5z_i^2}$ is the density of the standard normal distribution (with zero mean and SD equal to 1).

4. For each interval, calculate

$$\chi_i^2 = \frac{(\hat{n}_i - n_i)^2}{n_i},$$ (4.7)

and then, the statistic

$$\chi^2 = \sum_{i=1}^{l} \chi_i^2.$$ (4.8)

5. The number of degrees of freedom is defined as $k = l - 3$.
6. After selecting a confidence level q for the test, find χ_l^2 and χ_h^2 from the χ^2 distribution tables, satisfying the inequalities $P\{\chi^2 > \chi_l^2\} = 1 - 0.5q$ and $P\{\chi^2 > \chi_h^2\} = 0.5q$. The hypothesis regarding normal distribution is consistent with the experimental data if

$$\chi_l^2 < \chi^2 < \chi_h^2.$$ (4.9)

Example 4.2. To determine the density of distilled water at $T = 20°C$, 23 independent measurements were conducted. The results of the measurements came to $\rho_i = (0.998 + x_i \times 10^{-6})$ kg/l, where x_i are the values presented in Table 10.

Table 11 presents the test of the hypothesis of normal distribution of the errors of these measurements.

Table 10 Results of measurements of the density of distilled water

i	1	2	3	4	5	6	7	8	9
x_i	205	191	207	208	196	201	191	195	206
i	10	11	12	13	14	15	16	17	18
x_i	204	200	203	208	201	205	198	196	191
i	19	20	21	22	23	24	25		
x_i	193	195	197	192	196	194	198		

The mean of this sample comes to $\bar{x} = \frac{1}{25} \sum_{i=1}^{25} x_i = 198.8$

Table 11 Testing the hypothesis of the normal error distribution

i	Interval	$\hat{n}_i$	$\hat{x}_i$	z_i	n_i	χ_i^2
1	190–193	5	191.5	−1.24	2.467	2.602
2	193–196	6	194.5	−0.71	4.163	0.810
3	196–199	3	197.5	−0.17	5.271	0.978
4	199–202	3	200.5	0.36	5.002	0.803
5	202–205	3	203.5	0.90	3.564	0.089
6	205–208	5	206.5	1.44	1.904	5.034
	$\bar{x} = 198.5; S = 5.6$					$\chi^2 = 10.317$

As can be seen from the table, χ^2 statistic is 10.317. The number of degrees of freedom is $k = 6 - 3 = 3$. Let us take the confidence levels as $q_l = 97.5\%$ and $q_h = 2.5\%$, which ensures a confidence probability of 0.95. Using these and the number of degrees of freedom $k = 3$, we obtain: $\chi_{\mathcal{H}}^2 = 0.216$; $\chi_B^2 = 9.348$ from χ^2 distribution tables. Since $\chi^2 > \chi_B^2$, we come to the conclusion that the hypothesis of the normal distribution of the measurement errors is not compatible with the test data. Consequently, it is necessary to test the hypothesis regarding the generalized distribution of measurement errors. This procedure will be done in Sect. 4.4.

4.3 Direct Multiple Measurements Subject to Normal Distribution

With multiple measurements, the result is found by statistical analysis of a series of test data, frequently called parallel measurements. Now, a set of test data $y_1, ..., y_n$ is obtained in a measurement. A measurement result y_i might contain not only random error but also systematic error, and hence is called an uncorrected measurement result. An attempt is made to exclude systematic error by using specialized approaches. If it seems possible to estimate the systematic error Δ_c, then an adjustment is made to the test data, equal to the estimate of systematic error but taken with reverse sign:

$$x_i = y_i - \Delta_c.$$

The values of the x_i are called the corrected measurement results. They are burdened with random error and the residual portion of systematic error (non-excluded systematic error). Random error is estimated with statistical methods and hence defines the type A uncertainty of the measurement result. Non-excluded systematic error, estimated on the basis of a priori information, defines the type B uncertainty.

Statistical analysis is done as follows.

1. The mean of the test data is taken after the measurement result

$$\bar{x} = \frac{1}{n} \sum_{i=1}^{n} x_i, \tag{4.10}$$

where x_i is the uncorrected result of the ith measurement,
n is the number of measurements.

For computational convenience, this formula can be written as

$$\bar{x} = x_0 + \frac{1}{n} \sum_{i=1}^{n} z_i, \tag{4.11}$$

where x_0 is a value close to $\bar{x}$ that is convenient for calculation,

$$z_i = x_i - x_0.$$

2. The type A standard uncertainty of series of measurements is defined by the formula

$$u_A(x) = \sqrt{\frac{1}{n-1} \sum_{i=1}^{n} (x_i - \bar{x})^2} \quad \text{or} \quad u_A(x) = \sqrt{\frac{1}{n-1} \sum_{i=1}^{n} (z_i - \bar{z})^2}, \tag{4.12}$$

where

$$\bar{z} = \bar{x} - x_0 = \frac{1}{n} \sum_{i=1}^{n} z_i.$$

Dividing the sum in this formula by $n-1$, and not by n, is linked to the necessity of obtaining an unbiased estimate of the SD of the dispersal of the test data relative to the center of the distribution.

This formula can also be written as

$$u_A(x) = \sqrt{\frac{1}{n-1} \left[\sum_{i=1}^{n} x_i^2 - n(\bar{x})^2 \right]} \quad \text{or} \quad u_A(x) = \sqrt{\frac{1}{n-1} \left[\sum_{i=1}^{n} z_i^2 - n(\bar{z})^2 \right]}. \tag{4.13}$$

3. The type A standard uncertainty of the measurement result is determined by the
   formula

$$u_A(\bar{x}) = \frac{u_A(x)}{\sqrt{n}}. \tag{4.14}$$

4. All components of the type B uncertainty of the measurement result are
   estimated. The jth component (non-excluded systematic error caused by the jth
   factor) is estimated in the interval $[-U_j, +U_j]$. For each component, starting with
   assumptions about the form of its distribution and coverage probability P,
   the coverage factor k_j is determined.
5. All type B standard uncertainties are estimated using the formula

$$u_{Bj}(x) = \frac{U_j}{k_j}. \tag{4.15}$$

   If the jth component has an equal probability distribution in the interval
   $[-U_j, +U_j]$, then the standard uncertainty is found by the formula $u_{Bj}(x) = \frac{U_j}{\sqrt{3}}$.
6. The total standard uncertainty of the measurement result is estimated using the
   formula

$$u(\bar{x}) = \sqrt{u_A^2(\bar{x}) + \sum_{j=1}^{m} u_{Bj}^2(x)}. \tag{4.16}$$

7. The extended uncertainty of the measurement result is estimated in the interval
   $[-U(\bar{x}), +U(\bar{x})]$, the boundaries of which are calculated from the formula

$$U(\bar{x}) = k_P u(\bar{x}), \tag{4.17}$$

   in which k_P is the coverage factor of the uncertainty distribution, taken as equal
   to the quantile of Student's distribution $k_P = t_P(\nu_{\text{eff}})$ for coverage probability P
   and effective number of degrees of freedom of uncertainty distribution ν_{eff},
   calculated as

$$\nu_{\text{eff}} = (n-1) \times \frac{u^4(\bar{x})}{u_A^4(\bar{x})}. \tag{4.18}$$

Example 4.3. Calibrating a state standard reference sample (SSRS) of the mole
fraction of CO in nitrogen in a pressurized vessel.

The nominal value of the SSRS is $x_{SR} = 40$ ppm. The results of 14
measurements of this value are shown in Table 12.

Table 12   Results of measuring the SSRS value of the mole fraction of CO in nitrogen

i	1	2	3	4	5	6	7
x_i, ppm	45.00	36.25	42.50	45.00	37.50	38.33	37.50
i	8	9	10	11	12	13	14
x_i, ppm	43.33	40.53	36.25	42.50	39.17	45.00	40.83

Components of standard uncertainty evaluated as type B have the following estimates: $u_{B1}(x_{SR}) = 0.2\,\mathrm{ppm}$, $u_{B2}(x_{SR}) = 0.5\,\mathrm{ppm}$, $u_{B3}(x_{SR}) = 0.05\,\mathrm{ppm}$,

$$u_{B4}(x_{SR}) = 0.033 \times \bar{x}_{SR}\ \mathrm{ppm}.$$

The measurement result is determined by formula (4.10):
$\bar{x}_{SR} = \frac{1}{14}\sum\limits_{i=1}^{14} x_i = \frac{569.78}{14} = 40.70$ ppm. This value is assigned to the SSRS. Let us find its uncertainty. The standard uncertainty evaluated as type A is, in accordance with formulas (4.12) and (4.14), equal to

$$u_A(\bar{x}_{SR}) = \sqrt{\frac{1}{13 \times 14}\sum_{i=1}^{14}(x_i - \bar{x}_{CO})^2} = \sqrt{\frac{136.93}{13 \times 14}} \cong 0.87\ \mathrm{ppm}.$$

The total standard uncertainty of the measurement result is, in accordance with formula (4.16), equal to $u(\bar{x}_{SR}) = \sqrt{0.87^2 + 0.2^2 + 0.5^2 + 0.05^2 + (0.02 \times 40.7)^2} = 1.31$ ppm.

Let us estimate the extended uncertainty of the measurement result. The effective number of degrees of freedom is found by using formula (4.18): $v_{eff} = 13 \times \frac{1.31^4}{0.87^4} = 66.8$. For each value, we find from Table 5 the value of the coverage factor $k_{0.95} = t_{0.95}(66.8) = 2.05$. And, finally, by formula (4.17) we calculate the extended uncertainty of the measurement result: $U(\bar{x}_{SR}) = 2.05 \times 1.31 = 2.7$ ppm.

Hence, the actual value of the SSRS's mole fraction of CO in nitrogen is located in the interval [38.0–43.4] ppm with a coverage probability of 0.95.

Example 4.4. Conducting multiple measurements of the concentration of atmospheric components using a mass spectrometer.[1]

The mass spectrometer measurement method is based on the most universal property of matter – the difference in the masses of its component parts, atoms and molecules. The essence of this method lies in converting the atoms and molecules of the mixture being analyzed into charged particles – ions; dividing the ions into separate fractions, in each of which will be represented only ions of the same mass number $M = m/e$, where m and e are the mass and charge of the ion, respectively; and calculating the number of ions in each fraction.

---

[1] Laboratory work of the Department of Information Systems for Environmental Safety, SPbGPU.

Table 13 Mass spectrometer readings (collector current in mA)

i	1	2	3	4	5	6	7	8	9
x_i	390	379	344	422	368	376	368	383	397
i	10	11	12	13	14	15	16	17	18
x_i	393	405	407	385	395	415	397	393	371
i	19	20	21	22	23	24	25	26	27
x_i	410	384	366	390	378	395	390	431	367
i	28	29	30	31	32	33	3	35	36
x_i	405	336	432	405	402	396	412	396	410
i	37	38	39	40	41	42	43	44	45
x_i	356	419	412	432	393	390	393	349	432
i	46	47	48	49	50	51	52	53	54
x_i	427	398	397	406	410	405	401	415	413
i	55	56	57	58	59	60	61	62	
x_i	417	363	434	412	357	398	415	419	

One measure of the number of ions in one fraction is the value of the constant electric current excited in the collector circuit. Table 13 shows the results x_i of measuring the collector current.

The mean of this sample is

$$\bar{x} = \frac{1}{62} \sum_{i=1}^{62} x_i = 396\,\text{mA},$$

and the standard deviation of the set of measurements is

$$u_A(x) = \sqrt{\frac{1}{61} \sum_{i=1}^{62} (x_i - 396)^2} = 22.7\,\text{mA}.$$

Let us test the set of measurements for the presence of outliers. $x_{\text{max}} = 434\,\text{MA}$, $x_{\text{min}} = 336\,\text{MA}$. Then

$$G_{63} = \frac{434 - 396}{22.7} = 1.674, \quad G_1 = \frac{396 - 336}{22.7} = 2.643.$$

Since these values are less than the threshold value at a significance level $q = 0.05$, equal to 3.036 (see Table 6), the set of measurements does not contain outliers.

Let us test the hypothesis regarding normal error distribution of these measurements. In accordance with Table 7, for $n \cong 60$ we take $l = 7$ as the number of intervals. The width of an interval is $h = (434 - 336)/7 \cong 14$. We establish the boundaries of the intervals at: 336; 350; 364; 378; 392; 406; 420; and 434 mA. Further checking is shown in Table 14.

Table 14 Testing the hypothesis of the normal distribution of the set of measurements

i	Interval	$\hat{n}_i$	$\hat{x}_i$, MA	z_i	n_i	χ_i^2
1	336–350	3	343	−2.34	0.99	4.095
2	350–364	3	357	−1.72	3.48	0.067
3	364–378	7	371	−1.11	8.25	0.191
4	378–392	8	385	−0.49	3.55	2.274
5	392–406	20	399	0.13	15.15	1.550
6	406–420	14	413	0.75	11.53	0.528
7	420–434	7	427	1.36	6.06	0.146

The statistic $\chi^2 = \sum_{i=1}^{l} \chi_i^2 = 8.85$, and the number of degrees of freedom is $k = l - 3 = 7 - 3 = 4$. From χ^2 distribution tables for significance level $q = 0$ and $k = 4$ we find $\chi_l^2 = 0.872$ and $\chi_h^2 = 16.81$. Since $0.872 < 8.85 < 16.81$, the hypothesis of normal distribution is accepted.

The type A standard uncertainty for the result of multiple measurements is equal to $u_A(\bar{x}) = \frac{u_A(x)}{\sqrt{n}} = \frac{22.7}{\sqrt{62}} = 2.88\,\text{mA}$. Since the type B uncertainty is unknown, we shall consider that the total standard uncertainty is $u(\bar{x}) = u_A(\bar{x}) = 2.88\,\text{mA}$. The extended uncertainty is equal to $U(\bar{x}) = k \times u(\bar{x}) = 1.96 \times 2.88 = 5.54$, since, in accordance with Table 5, $k = t_{0.95}(v_{\text{eff}}) = 1.96$ for $v_{\text{eff}} = n - 1 = 61$.

The final result for the multiple measurements: the ion current of the mass spectrometer lies within the boundaries $[390 - 402]\,\text{mA}$ with a coverage factor of $k = 1.96$, corresponding to Student's distribution and a coverage probability $P = 0.95$.

4.4   Direct Multiple Measurements Subject to Generalized Normal Distribution

1. The base value of a measurand x_0 is, from the formula for the mean,

$$x_0 = \frac{1}{n} \sum_{i=1}^{n} x_i, \tag{4.19}$$

where x_i is the corrected result of the ith measurement,
n is the number of measurements.

If the distribution law for the random measurement error were normal, then the base value would be the conventional true value for the measurand. Since it has been established, by testing the hypothesis of normality, that the distribution law differs from the normal, actions shall be initiated that are directed toward refining this estimate, based on the generalized normal distribution.

2. From the formula

$$y_i = x_i - x_0 \tag{4.20}$$

the values y_i, $i = 1, ..., n$, of a series are defined, for which a generalized normal distribution will be selected using a yet unknown parameter p.

Since the generalized normal distribution law for values y_i corresponds to the normal distribution law for values $z_i = \text{sign}(y_i)(|y_i|)^p$, statistical analysis of the measurement results consists of the sequential execution of the following procedures:

- conversion from the series $y_1, ..., y_n$ to the series $z_1, ..., z_n$;
- calculation of the mean $\bar{z}$ of the series $z_1, ..., z_n$ and its standard and extended uncertainty, following the rules of statistical analysis of normally distributed test data;
- reverse conversion of the series $z_1, ..., z_n$ to the series $y_1, ..., y_n$
- calculation of the measurement result and its extended uncertainty.

3. An estimate $\hat{F}$ of the parameter F is found by the maximum likelihood method, which ensures the effectiveness of the statistical estimates derived. Determination of the unknown parameter of the distribution law of the random error for sample of measurement results $y_1, ..., y_n$ using the maximum likelihood method consist in finding a value of this parameter that renders specifically the appearance of this sample as the one most likely. In other words, the unknown parameter of the distribution is found using the condition for the maximum of the probability of obtaining the sample $y_1, ..., y_n$. Since all n measurement results are mutually independent, the likelihood function has the form

$$L(y_1, ..., y_n) = \prod_{i=1}^{n} f(y_i), \tag{4.21}$$

where $f(y_i)$ is the density of the generalized normal distribution, defined in (2.20). The condition of the maximum of the likelihood function $L(p, y_1, ..., y_n) = \max$ is transformed into an equation relative to the unknown F [15]:

$$\omega(F) = \ln\left(\frac{u_A(z)}{|F|}\right) - \frac{F-1}{n} \sum_{i=1}^{n} \ln(|y_i|) = \min, \tag{4.22}$$

in which

$$u_A(z) = \sqrt{\frac{1}{n-1} \sum_{i=1}^{n} (z_i - \bar{z})^2}$$

is the type A standard uncertainty of the set of values $z_i = \text{sign}(y_i)(|y_i|)^F$, and

$$\bar{z} = \frac{1}{n} \sum_{i=1}^{n} z_i$$

is the mean of this set.

Equation (4.22), as seen from the example for this section, can be easily solved with numerical methods.

The value $\hat{F}$ found from this equation is the statistical estimate of the parameter F. It was shown in [26] that the type A relative standard uncertainty of this estimate is equal to

$$u_{\text{rel}}(\hat{F}) = \frac{1}{2\sqrt{n}}. \tag{4.23}$$

The distribution law for the dispersal of the estimate $\hat{F}$ is unknown, and hence it is not possible to determine its extended uncertainty. Hence we shall employ the standard method recommended by the Guide: take the coverage factor $k = 2$, corresponding to a coverage probability of 0.95. Then, with a probability of 0.95, the true value of the parameter F will satisfy the following inequality:

$$\left(1 - \frac{1}{\sqrt{n}}\right)\hat{F} \leq F \leq \left(1 + \frac{1}{\sqrt{n}}\right)\hat{F}.$$

After F is found, the series $z_i = \text{sign}(y_i)(|y_i|)^F$, $i = 1, ..., n$ is formulated. After this, it is recommended to conduct an additional check of the agreement of the hypothesis regarding the generalized normal distribution with the test data. This is done by testing the hypothesis of a normal distribution of the series z_i (see Sect. 4.2).

4. The statistical characteristics of the series $z_1, ..., z_n$ are calculated, mean $\bar{z} = \frac{1}{n}\sum_{i=1}^{n} z_i$ and standard uncertainty $u_A(z) = \sqrt{\frac{1}{n-1}\sum_{i=1}^{n}(z_i - \bar{z})^2}$.

5. The measurement result is calculated. This is equal to the base value x_0, refined by introducing an adjustment $\bar{y} = \text{sign}(\bar{z}) \times (|\bar{z}|)^{1/F}$. In this way, the measurement result is expressed by the formula

$$x^* = x_0 + \text{sign}(\bar{z}) \times (|\bar{z}|)^{1/F}. \tag{4.24}$$

6. The type A standard uncertainty of the value of $\bar{z}$ is determined from the formula

$$u_A(\bar{z}) = \frac{u_A(z)}{\sqrt{n}}. \tag{4.25}$$

7. All components of the type B extended uncertainty of the measurement results x_i (and consequently of the members of the series y_i) are evaluated. The jth component is evaluated in the interval $[-U_j, +U_j]$. For each component, the coverage factor k_j is determined, proceeding from suppositions on the form of its distribution and the coverage probability P.

8. The boundaries are determined for all components of the type B extended uncertainty of the series z_i, corresponding to the boundaries of the uncertainty $\pm U_j$ of the series y_i:

$$\pm U_{Bj}(z) = \pm U_j^F. \tag{4.26}$$

9. The type B standard uncertainties of the series z_i are determined. For a distribution law of type B uncertainty for values y_i that differs from equal probability, one may use for this formula (3.6) and other recommendations of Sect. 3.3.
   If the type B uncertainty for values y_i has the equal probability distribution

$$f_B(\xi) = \begin{cases} \frac{1}{2a}, & -a \leq \xi \leq a, \\ 0, & \xi < -a, \ \xi > a \end{cases}$$

one may derive a precise estimate. We find for this the mean m_η and dispersion D_η of the distribution of the type B uncertainty η of the values z_i. In accordance with the rules of probability theory, these characteristics of the distribution of the random quantity $\eta = g(\xi) = \text{sign}(\xi) \times (|\xi|^F)$ are calculated as follows:

$$m_\eta = \int_{-\infty}^{\infty} g(\xi) f_B(\xi)\, d\xi = \int_{-a}^{a} \text{sign}(\xi)\, (|\xi|)^F \frac{1}{2a}\, d\xi$$

$$= \int_0^a \xi^F \frac{1}{2a}\, d\xi + \left[-\int_a^0 (-\xi^F)\frac{1}{2a}\, d\xi \right] = 0,$$

$$D_\eta = \int_{-\infty}^{\infty} [g(\xi) - m_\eta]^2 f_B(\xi) d\xi = \int_{-a}^{a} \text{sign}(\xi)\, (|\xi|)^{2F} \frac{1}{2a} d\xi$$

$$= \frac{a^{2F+1} - (-a^{2F+1})}{2a(2F+1)} = \frac{a^{2F}}{2F+1}.$$

In this way, the type B standard uncertainty of the values z_i, which is equal to the SD of this distribution, is expressed by the formula $u_B(z_i) = a^F/\sqrt{2F+1}$. It follows from this formula that for an equal probability distribution of the jth component of the uncertainty of the series y_i the corresponding standard uncertainties of the values z_i are calculated with the formula

$$u_{Bj}(z) = \frac{U_{Bj}(z)}{\sqrt{2F+1}}. \tag{4.27}$$

10. The total standard uncertainty of the value $\bar{z}$ is estimated by the formula

$$u(\bar{z}) = \sqrt{u_A^2(\bar{z}) + \sum_{j=1}^{m} u_{Bj}^2(z_i)}. \tag{4.28}$$

The total extended uncertainty of the value $\bar{z}$ is estimated as

$$U(\bar{z}) = k_P \times u(\bar{z}),\qquad(4.29)$$

where k_P is as in (4.18).

11. Now it is necessary to transition from the limits $\pm U(\bar{z})$ of the value $\bar{z}$ to the lower and upper bounds $U^-(\bar{y})$ and $U^+(\bar{y})$ of the extended uncertainty of the adjustment $\bar{y}$. They are equal to

$$U^-(\bar{y}) = \text{sign}[\bar{z} - U(\bar{z})][|\bar{z} - U(\bar{z})|]^{1/F},$$
$$U^+(\bar{y}) = \text{sign}[\bar{z} + U(\bar{z})][|\bar{z} + U(\bar{z})|]^{1/F}.\qquad(4.30)$$

Adding to these bounds the base value of the quantity, we derive the interval of uncertainty of the measurement result:

$$\left\{ x_0 + \text{sign}[\bar{z} - U(\bar{z})] \times [|\bar{z} - U(\bar{z})|]^{1/F},\ x_0 + \text{sign}[\bar{z} - U(\bar{z})] \times [|\bar{z} - U(\bar{z})|]^{1/F} \right\}.$$
$$(4.31)$$

This formula shows that, in contrast with the normal distribution, the interval of uncertainty is not symmetrical with respect to the measurement result $x^* = x_0 + \text{sign}(\bar{z}) \times (|\bar{z}|)^{1/F}$.

Example 4.5. Analysis of the test results in determining the density of distilled water at $T = 20°C$ (see Example 4.2)

The results of 25 independent measurements are presented in the form $\rho_i = (0.998 + x_i \times 10^{-6})\,\text{kg/l}$, where x_i are the values presented in Table 10 and the first column of Table 15. In the example mentioned, the mean value of the density $\rho_0 = (0.998 + x_0 \times 10^{-6})\,\text{kg/l}$ was found, where $x_0 = 198.8$. The calculations that were proposed in this example had as a goal the refinement of this result. With this goal in mind, ρ_0 is taken as the support value of the desired quantity, and the deviations $y_i \times 10^{-6}\text{kg/l}$ of the measurement results from this support value are found by formula (4.20). The values of y_i are shown in the second column of Table 15. The statistical check conducted in Sect. 4.2 has shown that the hypothesis of the normal distribution of the measurement error is rejected, at a significance level of the criterion at $q_B = 2.5\,\%$. Let us now check the possibility of accepting the hypothesis regarding the generalized normal distribution. We find the value of the parameter F of the distribution of values y_i that satisfies the condition of maximum likelihood, i.e., the condition of the minimum of the function $\omega(F)$ (4.22). We shall seek this value in the interval from 0.5 to 2.0 with a step of 0.1. The calculations run on an Excel spreadsheet are presented in Table 15.

The values of $\omega(F)$ are shown in the last line of this table. It is clear that the minimum of this function is reached at $F = 1.5$. The extended uncertainty of this value at a coverage factor of $k = 2$ is equal to $\left[\left(1 - \frac{1}{\sqrt{25}}\right) \times 1.6;\right.$ $\left.\left(1 + \frac{1}{\sqrt{25}}\right) \times 1.6\right] = [1.28;\ 1.92]$. The fact that this interval does not include the

4.4 Direct Multiple Measurements Subject to Generalized Normal Distribution

Table 15 Determining the parameter F

x_i	y_i	$\ln(|y_i|)$	$z_i = \text{sign}(y_i) \times (|y_i|)^F$ for F equal to															
			0.5	0.6	0.7	0.8	0.9	1	1.1	1.2	1.3	1.4	1.5	1.6	1.7	1.8	1.9	2.0
205	6.16	1.82	2.48	2.98	3.57	4.28	5.14	6.16	7.39	8.86	10.63	12.75	15.29	18.34	21.99	26.38	31.54	37.95
191	−7.84	2.06	−2.80	−3.44	−4.23	−5.19	−6.38	−7.84	−9.63	−11.84	−14.54	−17.87	−21.95	−26.97	−33.14	−40.72	−50.03	−61.64
207	8.16	2.10	2.86	3.52	4.35	5.36	6.61	8.16	10.07	12.42	15.32	18.9	23.31	28.75	35.47	43.76	53.98	66.59
208	9.16	2.21	3.03	3.78	4.71	5.88	7.34	9.16	11.43	14.27	17.8	22.22	27.72	34.6	43.17	53.88	67.24	83.91
196	−2.84	1.04	−1.69	−1.87	−2.08	−2.30	−2.56	−2.84	−3.15	−3.50	−3.88	−4.31	−4.79	−5.31	−5.90	−6.55	−7.27	−8.07
201	2.16	0.77	1.47	1.59	1.71	1.85	2.00	2.16	2.33	2.52	2.72	2.94	3.17	3.43	3.70	4.00	4.32	4.67
191	−7.84	2.06	−2.80	−3.44	−4.23	−5.19	−6.38	−7.84	−9.63	−11.84	−14.54	−17.87	−21.95	−26.97	−33.14	−40.72	−50.03	−61.47
195	−3.84	1.35	−1.96	−2.24	−2.56	−2.93	−3.38	−3.84	−4.39	−5.03	−5.75	−6.58	−7.52	−8.61	−9.85	−11.27	−12.89	−14.75
206	7.16	1.97	2.68	3.26	3.97	4.83	5.88	7.16	8.72	10.61	12.92	15.74	19.16	23.33	28.4	34.58	42.11	51.27
204	5.16	1.64	2.27	2.58	3.15	3.72	4.38	5.16	6.08	7.16	8.44	9.95	11.72	13.81	16.27	19.18	22.60	26.63
200	1.16	0.15	1.08	1.09	1.11	1.13	1.14	1.16	1.18	1.19	1.21	1.23	1.25	1.27	1.29	1.31	1.33	1.35
203	4.16	1.43	2.04	2.35	2.71	3.13	3.61	4.16	4.80	5.53	6.38	7.36	8.48	9.78	11.28	13.01	15.01	17.31
208	9.16	2.21	3.03	3.78	4.71	5.88	7.34	9.16	11.43	14.27	17.8	22.22	27.72	34.6	43.17	53.88	67.24	83.91
201	2.16	0.77	1.47	1.59	1.71	1.85	2.00	2.16	2.33	2.52	2.72	2.94	3.17	3.43	3.70	4.00	4.32	4.57
205	6.16	1.82	2.48	2.98	3.57	4.28	5.14	6.16	7.39	8.86	10.63	12.75	15.29	18.34	21.99	26.38	31.64	37.95
198	−0.84	−0.17	−0.92	−0.90	−0.89	−0.87	−0.85	−0.84	−0.83	−0.81	−0.80	−0.78	−0.77	−0.76	−0.74	−0.73	−0.72	−0.71
196	−2.84	1.04	−1.69	−1.87	−2.08	−2.30	−2.56	−2.84	−3.15	−3.50	−3.88	−4.31	−4.79	−5.31	−5.90	−6.55	−7.27	−8.07
191	−7.84	2.06	−2.80	−3.44	−4.23	−5.19	−6.38	−7.84	−9.63	−11.84	−14.54	−17.87	−21.95	−26.97	−33.14	−40.72	−50.03	−61.47
193	−5.84	1.76	−2.42	−2.88	−3.44	−4.10	−4.90	−5.84	−6.97	−8.31	−9.92	−11.83	−14.11	−16.84	−20.09	−23.96	−28.59	−34.11
195	−3.84	1.35	−1.96	−2.24	−2.56	−2.93	−3.36	−3.84	−4.39	−5.03	−5.75	−6.58	−7.52	−8.61	−9.85	−11.27	−12.89	−14.75
197	−1.84	0.61	−1.36	−1.44	−1.53	−1.63	−1.73	−1.84	−1.96	−2.08	−2.21	−2.35	−2.50	−2.65	−2.82	−3.00	−3.19	−3.39
192	−6.84	1.92	−2.62	−3.17	−3.84	−4.66	−5.64	−6.84	−8.29	−10.05	−12.18	−14.76	−17.89	−21.68	−26.28	−31.85	−38.6	−46.79
196	−2.84	1.04	−1.69	−1.87	−2.08	−2.30	−2.56	−2.84	−3.15	−3.50	−3.88	−4.31	−4.79	−5.31	−5.90	−6.55	−7.27	−8.07
194	−4.84	1.58	−2.20	−2.58	−3.02	−3.53	−4.13	−4.84	−5.67	−6.63	−7.77	−9.09	−10.65	−12.47	−14.6	−17.09	−20.01	−23.43
198	−0.84	−0.17	−0.92	−0.90	−0.89	−0.87	−0.85	−0.84	−0.83	−0.81	−0.80	−0.78	−0.77	−0.76	−0.74	−0.73	−0.72	−0.71
$\bar{x} = 198.8$		$u_A(z_p) =$	2.25	2.68	3.21	3.87	4.66	5.64	6.84	8.30	10.10	12.30	15.01	18.33	22.41	27.44	33.61	41.21
		$\omega(F) =$	2.191	2.048	1.937	1.851	1.783	1.730	1.689	1.659	1.637	1.623	1.615	**1.613**	1.615	1.623	1.634	1.649

Table 16 Testing the hypothesis regarding the generalized normal distribution of values y_i

i	Interval of values z_i	$\hat{n}_i$	z_{i0}	$\xi_{i0} = \frac{z_{i0}-\bar{z}}{S(z)}$	n_i	χ_i^2
1	2	3	4	5	6	7
1	−28.0 to (−17.5)	5	−22.75	−1.391	2.396	1.074
2	−17.5 to (−7.0)	6	−12.25	−0.759	4.729	0.112
3	−7.0 to 3.5	3	−1.75	−0.126	6.258	1.202
4	3.5 to 14.0	3	8.75	0.506	5.550	2.271
5	14.0 to 24.5	3	19.25	1.138	3.300	0.027
6	24.5 to 35.0	5	29.75	1.717	1.315	2.159
	$\bar{z}=0.35$; $S(z) = 16.60$					$\chi^2 = 6.845$

value $F = 1$, corresponding to the normal distribution, serves as additional confirmation as to the validity of the decision to reject the hypothesis.

In any case, we check with χ^2 test the hypothesis of the generalized normal distribution of values y_i with parameter $F = 1.5$. For this, it is sufficient to be convinced of the normal distribution of values $z_i = \text{sign}(y_i) \times (|y_i|)^{1.6}$. We conduct the test in tabular form, analogous to Example 4.2. As is clear from Table 15, the minimum value of z_i is equal to -27.0, and the maximum value is 34.5. Let us take $z_{\min} = -28$, $z_{\max} = 35$, and number of intervals $n = 6$. The width of an interval is $h = (35 - (-28))/6 = 10.5$. These intervals are shown in the second column of Table 16. The third column shows the empirical frequencies $\hat{n}_i$ of incidence in the ith interval, the fourth shows the central points z_{i0} of these intervals, and the fifth shows the normalized values of these points $\xi_{i0} = (z_{i0} - \bar{z})/S(z)$.

The sixth column shows the theoretical values of the frequencies of incidence in these intervals, calculated from the formula $n_i = n(h/S(z))f(\xi_{i0})$, where $f(\xi_{i0})$ is the normal distribution density with zero mean and SD equal to 1. The last column shows the values $\chi_i^2 = (\hat{n}_i - n_i)^2/n_i$ and their sum, the statistic $\chi^2 = 6.845$.

As in the preceding example, the number of degrees of freedom $k = 3$. As earlier, we take $q_{\mathcal{H}} = 97.5\%$ and $q_B = 2.5\%$. For these and for $k = 3$ we obtain from the χ^2 distribution tables: $\chi_{\mathcal{H}}^2 = 0.216$; $\chi_B^2 = 9.348$. Since $0.216 < 6.845 < 9.348$, we conclude that the hypothesis of the generalized normal distribution of the random measurement error with parameter $F = 1.6$ is in agreement with the test data.

We find the statistical characteristics of the series z_i for the elements of the series, shown in Table 15: mean $\bar{z} = 0.82$ and unbiased standard deviation $S(z) = 18.33$. The type A standard uncertainty of the value $\bar{z}$ is consequently equal to $u_A(\bar{z}) = \frac{18.33}{\sqrt{25}} = 3.67$.

Let us find the type B standard uncertainty for the series z_i. The type B extended uncertainty of the measurement result for the density of distilled water is due to the imprecision of adjustment for the density of air and is estimated as 0.75×10^{-6} kg/l. Consequently, $U_B(y) = 0.75$. The extended uncertainty of the values z_i is equal to $U_B(z) = 0.75^{1.6} = 0.63$, and the standard uncertainty is $u_B(\bar{z}) = 0.63/\sqrt{2 \times 1.6 + 1} = 0.31$. Taking account of these estimates, the total standard uncertainty of the value $\bar{z}$ is equal to $u_c(\bar{z}) = \sqrt{3.67^2 + 0.31^2} = 3.68$.

The effective number of degrees of freedom is equal to

$$\nu_{\text{eff}} = 24 \times \frac{3.68^4}{3.67^4} \cong 24.$$

By Table 5, for this number of degrees of freedom and a coverage probability of $P = 0.95$, we find $t_{0.95}(24) = 2.12$. Consequently, the total extended uncertainty of $\bar{z}$ is $U(\bar{z}) = 2.12 \cdot 3.68 = 7.80$.

Now we can find the measurement result and its extended uncertainty $\bar{y} = \text{sign}(\bar{z}) \times (|\bar{z}|)^{1/\hat{F}} = 0.82^{1/1.6} = 0.9$.

$$\rho^* = \rho_0 + \bar{y} \times 10^{-6} = 0.998 + (x_0 + \bar{y}) \times 10^{-6}$$
$$= 0.998 + (198.8 + 0.9) \times 10^{-6} = 0.9981997\,\text{kg/l}.$$

The extended uncertainty $\bar{y}$ in accordance with (4.30) is equal to $-[7.80 - 0.82]^{1/1.6}$; $[7.80 + 0.82]^{1/1.6} = [-3.37;\ 3.84]$.

Consequently, the extended uncertainty of the measurement result is:

$$[\rho^-;\ \rho^+] = 0.998 + (198.8 + [-3.37;\ 3.84]) \times 10^{-6}$$
$$= 0.9981988 + [-3.37;\ 3.84] \times 10^{-6}$$
$$= [0.9981954;\ 0.9982026]\,\text{kg/l}.$$

And so, the measurement result for the density of distilled water at $T = 20°C$ is $\bar{\rho} = 0.9981997\,\text{kg/l}$. It is clear that the statistical analysis of the results of this experiment, based on the generalized normal distribution, permitted refining the value of this physical constant by one unit at the sixth decimal place. The estimate of the uncertainty of this measurement provides evidence that the true value of this constant is, with coverage probability 0.95, located in the interval [0.9981954; 0.9982026] kg/l.

4.5 Indirect Measurements

When conducting indirect measurements, the value of the quantity being sought is found by solving the measurement equation

$$y = f(x_1, ..., x_n), \tag{4.32}$$

which links the value

$$u_A\left(\sum_{j=1}^{4} v_j^2\right) = \sqrt{\frac{\sum_{j=1}^{4} v_j^2}{4 - 2}} = \sqrt{\frac{0.003}{2}} = 0.039\,\text{V}.$$

of the measurand Y with the values x_i of other quantities $X_1, ..., X_n$, which are directly measured or known, referred to in 3.3 as input quantities. From the form of the functional dependence, the indirect measurements are divided into two kinds – those with linear and those with nonlinear dependencies. The mathematical apparatus for statistical analysis of indirect measurements of the first type has been worked out in detail. There is no mathematically rigid method for a nonlinear dependence. For these kinds of problems, several approximation methods are used, the basic ones of which are linearization and reduction.

4.5.1 Statistical Manipulation with the Linear Dependence of Measurement Result on Input Quantities

In this case, the measurement equation has the form

$$y = \sum_{i=1}^{n} a_i x_i, \tag{4.33}$$

where a_i are constant coefficients of the quantities x_i, which are arguments of the equation, and n is the number of arguments.

If the coefficients a_i are also determined experimentally, then the measurement result is found in stages: first each term $a_i x_i$ is evaluated as the result of an indirect measurement of a quantity that is equal to the product of two other quantities; then the value $\bar{y}$ of the unknown quantity is calculated and its uncertainty is evaluated.

The result $\bar{y}$ of an indirect measurement is found as follows. The quantities $X_1, ..., X_n$ are measured, and adjustments for all known systematic errors are introduced into the results of these measurements. The estimates $\bar{x}_1, ..., \bar{x}_n$ of the measured quantities are determined. If the quantity X_i was measured once, then $\bar{x}_i = x_{i1}$, where x_{i1} is the result of a single measurement. If this quantity was measured multiple times, then

$$\bar{x}_i = \frac{1}{m_i} \sum_{j=1}^{m_i} x_{ij}, \tag{4.34}$$

where m_i is the number of measurements of the quantity X_i. Hereafter (4.34) shall be kept in mind, since the result of a single measurement is a particular case of this.

The result of an indirect measurement is calculated with the formula

$$\bar{y} = \sum_{i=1}^{n} a_i \bar{x}_i. \tag{4.35}$$

At the next stage, the standard uncertainties $u(x_i)$, $i = 1, ..., n$, of the arguments are calculated. If there are reasons to suppose there is correlation of the measurement results $\bar{x}_i$ and $\bar{x}_j$ of all or some of the arguments, then their correlation coefficients r_{ij} are calculated.

The procedure for estimating standard uncertainties is presented in Sect. 3.3. The type A standard uncertainty is calculated using the formula for the SD of the random error of the result of multiple measurements:

$$u_A(\bar{x}_i) = \sqrt{\frac{1}{m_i(m_i - 1)} \sum_{j=1}^{m_i} (x_{ij} - \bar{x}_i)^2}. \tag{4.36}$$

If the estimate of the quantity Y was derived by taking account of the result of one measurement of the quantity X_i, then another $(m_i - 1)$ of its measurements are brought in to estimate the uncertainty, and based on them $u(\bar{x}_i)$ is calculated using the formula for the SD of the random error of the result of a single measurement.

$$u_A(\bar{x}_i) = \sqrt{\frac{1}{(m_i - 1)} \sum_{j=1}^{m_i} (x_{ij} - \bar{x}_i)^2}. \tag{4.37}$$

The input data for evaluating type B standard uncertainty is enumerated in Sect. 3.3. The uncertainties of this type are usually presented in the form of bounds $\pm U_{Bj}(x_i)$, $j = 1, ..., l_i$, of the extended uncertainty, showing the coverage factor k_{ij} (here l_i is the number of sources of type B uncertainty of the measurement results $\bar{x}_i$). When the distribution law is unknown, the most widely used method is to presume an equal probability distribution law within the specified bounds. In this case, $k_{ij} = \sqrt{3}$.

The type B standard uncertainties are, in accordance with Sect. 3.3, calculated from the formula

$$u_{Bj}(\bar{x}_i) = \frac{U_{Bj}(x_i)}{k_{ij}}. \tag{4.38}$$

The standard uncertainty of the result $\bar{x}_i$ is equal to the mean square sum of these estimates:

$$u(\bar{x}_i) = \sqrt{u_A^2(\bar{x}_i) + \sum_{j=1}^{l_i} u_{Bj}^2(\bar{x}_i)}. \tag{4.39}$$

Correlation coefficients are estimated according to (3.11) and (3.12). The significance of correlation links between the estimates of the arguments is analyzed. The criterion for lack of such link if the following inequality is true:

$$\left| \frac{r(x_i, x_j)}{\sqrt{1 - r^2(x_i, x_j)}} \right| < \frac{t_P(m - 2)}{\sqrt{m - 2}}, \tag{4.40}$$

where $t_P(m-2)$ is the coefficient of Student's distribution for probability P and $(m-2)$ degrees of freedom, and

$$m = \min(m_i, \, m_j).$$

The total standard uncertainty $u(\bar{y})$ is calculated. For uncorrelated estimates $\bar{x}_1, ..., \bar{x}_n$, this is determined by the formula

$$u(\bar{y}) = \sqrt{\sum_{i=1}^{n} a_i^2 u^2(\bar{x}_i)},$$ (4.41)

and if there is correlation, by the formula

$$u(\bar{y}) = \sqrt{\sum_{i=1}^{n} a_i^2 u^2(\bar{x}_i) + \sum_{\substack{i,j=1 \\ i \neq j}}^{n} a_i a_j r(x_i, x_j) \bar{u}(x_i) \bar{u}(x_j)}.$$ (4.42)

The extended uncertainty of the measurement result is determined by formula (3.16), in which the coverage factor k is defined by (3.17), and the effective number of degrees of freedom v_{eff} is calculated using the formula

$$v_{\text{eff}} = \frac{u^4(\bar{y})}{\sum_{i=1}^{n} a_i^4 u^4(\bar{x}_i)/v_{i.\text{eff}}},$$ (4.43)

in which $v_{i.\text{eff}}$ is the effective number of degrees of freedom when the estimate $\bar{x}_i$ is determined.

If all type B uncertainties are evaluated as following the equal probability law, then for each of them $v_i = \infty$. Then the effective number of degrees of freedom for the estimate $\bar{x}_i$ is equal to

$$v_{i,\text{eff}} = (m_i - 1) \times \frac{u^4(\bar{x}_i)}{u_A^4(\bar{x}_i)},$$

where m_i is the number of measurements of the quantity $u(y) = \sqrt{u_A^2(\bar{y}) + u_B^2(y)} = \sqrt{0.035^2 + 0.03^2} = 0.046 \, \text{V/ppm}..$ Substituting this formula into (4.43), we derive, for this case:

$$v_{\text{eff}} = \frac{u^4(\bar{y})}{\sum_{i=1}^{m} a_i^4 u_A^4(\bar{x}_i)/m_i - 1}.$$ (4.44)

Now let us examine a nonlinear dependence $y = f(x_1, ..., x_n)$.

4.5.2 *Method of Linearization of Nonlinear Dependence*

This method presupposes the expansion of this function into a Taylor series

$$y = f(x_1, \ ..., \ x_n) = f(\bar{x}_1, \ ..., \ \bar{x}_n) + \sum_{i=1}^{n} \frac{\partial f(\bar{x}_1, \ ..., \ \bar{x}_n)}{\partial x_i} \Delta x_i + R, \qquad (4.45)$$

where $\partial f(\bar{x}_1, \ ..., \ \bar{x}_n)/\partial x_i$ is the first derivative of the function $f(x_1, \ ..., \ x_n)$ on x_i, calculated at the point $(\bar{x}_1, \ ..., \ \bar{x}_n)$, $\Delta x_i = x_i - \bar{x}_i$, and

$$R = \frac{1}{2} \left[\sum_{i=1}^{n} \frac{\partial^2 f(\bar{x}_1, \ ..., \ \bar{x}_n)}{\partial x_i^2} (\Delta x_i)^2 + \sum_{\substack{i,k = 1 \\ i \neq k}}^{n} \frac{\partial^2 f(\bar{x}_1, \ ..., \ \bar{x}_n)}{\partial x_i \partial x_k} \Delta x_i \Delta x_k \right]$$

is the residual term of the expansion.

The linearization method is feasible if the increment of the function $\Delta f(x_1, \ ..., \ x_n) = f(x_1, \ ..., \ x_n) - f(\bar{x}_1, \ ..., \ \bar{x}_n)$ can be replaced by its total differential $\sum_{i=1}^{n} (\partial f(\bar{x}_1, \ ..., \ \bar{x}_n)/\partial x_i) \Delta x_i$, ignoring the residual term. The condition for R to be insignificant is:

$$R < 0.8 u(\bar{y}). \qquad (4.46)$$

In checking this condition, the deviations Δx_i must be derived such that they are actually possible, and that in using them the maximum of the function $f(x_1, ..., x_n)$ is reached. Hence, with this method of execution, the solution reduces to a linear dependence of the measurand on the input quantities. The measurement result is taken as

$$\bar{y} = f(\bar{x}_1, \ ..., \ \bar{x}_n). \qquad (4.47)$$

The second term of formula (4.45) serves to estimate the uncertainty of this measurement result. The equation

$$\Delta y = \sum_{i=1}^{n} \frac{\partial f(\bar{x}_1, \ ..., \ \bar{x}_n)}{\partial x_i} \Delta x_i \qquad (4.48)$$

is an equation of the indirect measurements of the absolute measurement error Δy, which is determined from the known errors Δx_i. On the other hand, this equation of measurements pertains to the linear equations for which the correct solution method was examined above. Hence setting $a_i = \partial f(\bar{x}_1, ..., \bar{x}_n)/\partial x_i$ and taking as its quantities the absolute arguments of the measurement errors Δx_i, we derive, by analogy with formulas (4.41) and (4.42) for uncorrelated values of input quantities,

$$u(\bar{y}) = \sqrt{\sum_{i=1}^{n} \left[\frac{\partial f(\bar{x}_1, \ldots, \bar{x}_n)}{\partial x_i}\right]^2 u^2(\bar{x}_i)}, \qquad (4.49)$$

and when there is correlation:

$$u(\bar{y}) = \sqrt{\sum_{i=1}^{n} \left[\frac{\partial f(\bar{x}_1, \ldots, \bar{x}_n)}{\partial x_i}\right]^2 u^2(\bar{x}_i) + \sum_{\substack{i,j = 1 \\ i \neq j}}^{n} \frac{\partial f(\bar{x}_1, \ldots, \bar{x}_n)}{\partial x_i} \times \frac{\partial f(\bar{x}_1, \ldots, \bar{x}_n)}{\partial x_j} r(x_i, x_j) \bar{u}(x_i) \bar{u}(x_j)}.$$

$$(4.50)$$

The extended uncertainty of a measurement result is defined by formula (3.16), into which is substituted the coverage factor, determined by taking account of the effective number of degrees of freedom as calculated by formula (4.43). If all type B uncertainties are evaluated by the equal probability law, then the effective number of degrees of freedom is defined, in accordance with formula (4.44), as:

$$\nu_{\text{eff}} = \frac{u^4(\bar{y})}{\sum_{i=1}^{n} \left([\partial f(\bar{x}_1, \ldots, \bar{x}_n)/\partial x_i]^4 u_A^4(\bar{x}_i) \right)/m_i - 1}. \qquad (4.51)$$

Example 4.6. It is necessary to estimate the result of an indirect measurement of the density of a hard body, as well as its uncertainty.

We find the density of a hard body using the formula $\rho = M/V$, where M is the mass of the body and V is its volume. For this we perform 11 measurements of the mass and volume and calculate the measurement results of the input quantities:

$$m_M = m_V = m = 11, \quad \bar{M} = 252.9120 \text{ kg}, \quad \bar{V} = 195.3798 \times 10^{-3} \text{ m}^3,$$
$$u^2(\bar{M}) = 19.4 \times 10^{-14} \text{ kg}^2, u^2(\bar{V}) = 16.4 \times 10^{-20} \text{ m}^6,$$
$$\max(|\Delta M|) = 31 \times 10^{-7} \text{ kg}, \quad \max(|\Delta V|) = 32 \times 10^{-10} \text{ m}^3.$$

The measurement result is equal to

$$\bar{\rho} = \frac{\bar{M}}{\bar{V}} = \frac{252.9120}{195.3798 \times 10^{-3}} = 1.294463 \times 10^3 \frac{\text{kg}}{\text{m}^3}.$$

To estimate the uncertainty of this result, we use the linearization method. Let us find the derivatives.

$$\frac{\partial \rho(\bar{M}, \bar{V})}{\partial M} = \frac{\partial \rho(M, V)}{\partial M}\Big|_{M=\bar{M},\, V=\bar{V}} = \frac{1}{V}\Big|_{V=\bar{V}} = \frac{1}{\bar{V}},$$

$$\frac{\partial \rho(\bar{M}, \bar{V})}{\partial V} = \frac{\partial \rho(M, V)}{\partial V}\Big|_{M=\bar{M},\, V=\bar{V}} = -\frac{M}{V^2}\Big|_{M=\bar{M},\, V=\bar{V}} = -\frac{\bar{M}}{\bar{V}^2} = -\frac{\bar{\rho}}{\bar{V}},$$

$$\frac{\partial^2 \rho(\bar{M}, \bar{V})}{\partial M^2} = \frac{\partial^2 \rho(M, V)}{\partial M^2}\Big|_{M=\bar{M},\, V=\bar{V}} = \left[\frac{\partial}{\partial M}\left(\frac{1}{V}\right)\right]\Big|_{M=\bar{M},\, V=\bar{V}} = 0,$$

$$\frac{\partial^2 \rho(\bar{M}, \bar{V})}{\partial V^2} = \frac{\partial^2 \rho(M, V)}{\partial V^2}\Big|_{M=\bar{M},\, V=\bar{V}} = \left[\frac{\partial}{\partial V}\left(-\frac{M}{V^2}\right)\right]\Big|_{M=\bar{M},\, V=\bar{V}} = 2\frac{M}{V^3}\Big|_{M=\bar{M},\, V=\bar{V}}$$

$$= 2\frac{\bar{M}}{\bar{V}^3} = 2\frac{\bar{\rho}}{\bar{V}^2},$$

$$\frac{\partial^2 \rho(\bar{M}, \bar{V})}{\partial M \partial V} = \frac{\partial^2 \rho(M, V)}{\partial M \partial V}\Big|_{M=\bar{M},\, V=\bar{V}} = \left[\frac{\partial}{\partial V}\left(\frac{1}{V}\right)\right]\Big|_{V=\bar{V}} = -\frac{1}{V^2}\Big|_{V=\bar{V}} = -\frac{1}{\bar{V}^2}$$

The standard uncertainty of the measurement result is equal to

$$u(\rho) = \sqrt{\left[\frac{\partial \rho(\bar{M}, \bar{V})}{\partial M}\right]^2 u^2(\bar{M}) + \left[\frac{\partial \rho(\bar{M}, \bar{V})}{\partial V}\right]^2 u^2(\bar{V})} = \sqrt{\frac{u^2(\bar{M}) + \bar{\rho}^2 u^2(\bar{V})}{\bar{V}^2}}$$

$$= \sqrt{\frac{19.4 \times 10^{-14} + (1.294463 \times 10^3)^2 \times 16.4 \times 10^{-20}}{(195.3798 \times 10^{-3})^2}} = 3.5 \times 10^{-6}\ \frac{\text{kg}}{\text{m}^3}.$$

The residual term of the Taylor series expansion is equal to

$$R = \frac{1}{2}\left[\frac{\partial^2 \rho(\bar{M}, \bar{V})}{\partial M^2}(\Delta M)^2 + \frac{\partial^2 \rho(\bar{M}, \bar{V})}{\partial V^2}(\Delta V)^2 + 2\frac{\partial^2 \rho(\bar{M}, \bar{V})}{\partial M \partial V}\Delta M \Delta V\right]$$

$$= \frac{\bar{\rho}(\Delta V)^2 - \Delta M \Delta V}{\bar{V}^2}.$$

Since the increments ΔM and ΔV can be either positive or negative, we estimate R with the formula:

$$R = \frac{\bar{\rho}(\Delta V)^2 + \Delta M \Delta V}{\bar{V}^2}$$

$$= \frac{1.294463 \times 10^3 \times 32^2 \times 10^{-20} + 31 \times 10^{-7} \times 32 \times 10^{-10}}{195.3798^2 \times 10^{-6}}$$

$$= 6.07 \times 10^{-11}.$$

Since condition (4.46) is fulfilled: $(6.07 \times 10^{-11} \ll 0.8 \cdot 3.5 \times 10^{-6})$, use of the linearization method is permissible. Let us find the extended uncertainty of the measurement result. The effective number of degrees of freedom by formula (4.51) will be equal to

$$v_{\text{eff}} = \frac{u^4(\bar{\rho})}{[\partial\rho(\bar{M}, \ \bar{V})/\partial M]^4 u^4(\bar{M})/m_M - 1 + (\partial\rho(\bar{M}, \ \bar{V})/\partial V)^4 u^4(\bar{V})/m_V - 1}$$

$$= (m-1)\bar{V}^4 \frac{u^4(\bar{\rho})}{u^4(\bar{M}) + \bar{\rho}^4 u^4(\bar{V})}$$

$$= 10(195.3798 \times 10^{-3})^4 \frac{(3.5 \times 10^{-6})^4}{(19.4 \times 10^{-14})^2 + (1.294463 \times 10^3)^4 (16.4 \times 10^{-20})^2}$$

$$= 10\frac{(1.953798 \times 3.5)^4}{19.4^2 + 1.294463^4 \times 16.4^2} = \frac{10 \times 2186.7}{376.36 + 755.17} \cong 19.$$

The coverage factor k, in accordance with Table 5, will be equal to 2.14. The extended uncertainty of the measurement result is $U(\bar{\rho}) = 2.14 \times 3.5 \times 10^{-6} = 7.5 \times 10^{-6} \text{ kg/m}^3$. Consequently, the measurand, with a coverage factor of 2.14 corresponding to a coverage probability of 0.95, is located in the interval:

$$[(1.294463 - 7.5 \times 10^{-6}) - (1.294463 + 7.5 \times 10^{-6})]\frac{\text{kg}}{\text{m}^3}$$

$$= (1.294455 - 1.294470)\frac{\text{kg}}{\text{m}^3}.$$

4.5.3   Reduction Method [4]

This method of analyzing the results of indirect measurements is used when the use of the linearization method does not guarantee the required accuracy of a measurement result due to not fulfilling condition (4.46). It consists of presenting the set of values of the indirectly measured quantity in the form of a series of results of primary measurements. Different combinations of the separate measurement results x_{ij} of the input quantities X_i are substituted into the equation for indirect measurement (4.32), and the corresponding values y_s $(s = 1, ..., m)$ of the measurand Y are calculated. The series of values y_s thus obtained can be studied as a series of results of primary measurements. The measurement result is calculated by the formula $\bar{y} = 1/m \sum_{s=1}^{m} y_s$, the type A standard uncertainty by the formula

$$u_A(\bar{y}) = \sqrt{\frac{1}{m(m-1)} \sum_{s=1}^{m} (y_s - \bar{y})^2},$$

and the type A extended uncertainty by the formula $U_A(\bar{y}) = t_P(m-1)u_A(\bar{y})$, where $t_P(m-1)$ is the quantile of Student's distribution. Estimates of type B uncertainty and the total uncertainty can also be calculated by the usual method. However, it is recommended here not to forget that the latter were derived on the basis of using a series of assumptions that were loosened in practice in this case [in particular, condition (4.46)], and consequently are very approximate.

4.6 Joint and Aggregate Measurements

In measurement practice, cases are met in which the unknown quantities cannot be measured directly or represented as explicit functions of directly measured quantities. In such cases, quantities that are functionally related to the unknown quantities are measured, and the values of the latter are calculated using a system of implicit equations

$$F_j(x_{1j}, x_{2j}, ..., x_{nj}, y_1, y_2, ..., y_k) = 0, \quad j = 1, 2, ..., m, \qquad (4.52)$$

where F_j is the symbol for the functional dependence between the quantities in the jth test, $x_{1j}, x_{2j}, ..., x_{nj}$ are the measurement results of directly measured quantities $X_1, X_2, ..., X_n$ in the jth test, $y_1, y_2, ..., y_k$ are the unknown values of the measurands $Y_1, Y_2, ..., Y_k$.

If the $Y_1, Y_2, ..., Y_k$ and $X_1, X_2, ..., X_n$ are values of different nature, then the measurements described by (4.52), are called joint. If all these quantities are homogeneous, then the measurements are called aggregate.

After substituting into the initial system of equations the results x_{ij} of direct or indirect measurements, they have the form

$$F_j(y_1, y_2, ..., y_k) = 0, \quad j = 1, 2, ..., m. \qquad (4.53)$$

These equations are not an exact reflection of the true relationship between the unknown quantities $y_1, y_2, ..., y_k$, since they are burdened with the errors x_{ij} in the measurements results. Hence they are referred to as conditional. The general method for solving such systems of equations consists of the following. Among the infinite set of possible solutions of this system (and the number of solutions will be infinite for $L = 10 \times \lg_{10}(I/I_0),$) some best solution $\bar{y}_1, \bar{y}_2, ..., \bar{y}_k$ is found. If this solution is substituted into the conditional equations, then, due to the uncertainties of the measurement results x_{ij}, the right sides of the equations will differ from the left sides. To obtain identities, it is necessary to write

$$F_j(\bar{y}_1, \bar{y}_2, ..., \bar{y}_k) + v_j = 0, \quad j = 1, 2, ..., m, \qquad (4.54)$$

where v_j are the residual errors of the conditional equations and are called *residuals*.

In accordance with Legendre's principle, the equations being solved will have the most reliable solution if the sum of squares of all residuals is minimal:

$$\sum_{j=1}^{m} v_j^2 = \min. \tag{4.55}$$

The solution of conditional equations using Legendre's principle is called the method of least squares. It follows from (4.55) that the complete differential must equal zero:

$$d\left(\sum_{j=1}^{m} v_j^2\right) = \frac{\partial\left(\sum_{j=1}^{m} v_j^2\right)}{\partial \bar{y}_1} d\bar{y}_1 + \frac{\partial\left(\sum_{j=1}^{m} v_j^2\right)}{\partial \bar{y}_2} d\bar{y}_2 + \cdots + \frac{\partial\left(\sum_{j=1}^{m} v_j^2\right)}{\partial \bar{y}_k} d\bar{y}_k = 0.$$

In turn, this equality will be fulfilled for any values of the differentials $d\bar{y}_i$ only if the equations are valid:

$$\frac{\partial\left(\sum_{j=1}^{m} v_j^2\right)}{\partial \bar{y}_i} = 0, \quad i = 1, ..., k. \tag{4.56}$$

The system (4.56) is called a system of normal equations. It consists of k equations relative to k unknowns and provides a single solution of the unknown values $\bar{y}_i$ of the quantities Y_i.

In solving this problem in the general case, with nonlinear conditional equations and correlation of the results of separate measurements, a series of insurmountable difficulties arises. Hence, in practice, an attempt is always made, one way or another, to convert nonlinear conditional equations to linear form. One of the most widespread methods consists of replacing the unknowns in such a way that the conditional equations would be linear with respect to the new unknown quantities. Another method is to expand the nonlinear function into a Taylor series, eliminating the nonlinear part of the series.

As an example, let us examine the joint measurements of three diverse quantities – X, Y, and Z – based on a link equation

$$ax + by + cz = l, \tag{4.57}$$

in which a, b, c, and l are values of directly measured quantities (input quantities), and x, y, and z are the unknown values of the quantities X, Y, and Z (output quantities).

Substituting into (4.57) the measurement results a_i, b_i, c_i, l_i, we obtain a series of conditional equations:

$$a_1 x + b_1 y + c_1 z = l_1,$$
$$a_2 x + b_2 y + c_2 z = l_2,$$
$$\vdots$$
$$a_m x + b_m y + c_m z = l_m.$$

If the number of equations is less than the number of unknown values, the problem is unsolvable. If it is equal, a numerical value may be found for each quantity. However, they will be burdened with unknown errors due to the uncertainties of the measurements of the quantities a_j, b_j, c_j, l_j. Hence, the number of equations in this system must exceed the number of unknowns. If one takes account of the measurement error of the input quantities, then this system is written in this form:

$$a_1 x + b_1 y + c_1 z - l_1 + v_1 = 0,$$
$$a_2 x + b_2 y + c_2 z - l_2 + v_2 = 0,$$
$$\vdots$$
$$a_m x + b_m y + c_m z - l_m + v_m = 0,$$

where $v_1, v_2, \ldots, v_m$ are the residuals of the equations.

To compile normal equations, we consider that

$$\sum_{j=1}^{m} v_j^2 = \sum_{j=1}^{m} (a_j a_j x^2 + 2a_j b_j xy + 2a_j c_j xz - 2a_j l_j x + b_j b_j y^2 + 2b_j c_j yz - 2b_j l_j y$$
$$+ c_j c_j z^2 - 2c_j l_j z + l_j^2).$$

Consequently, the first equation has the form $\dfrac{\partial\left(\sum_{j=1}^{m} v_j^2\right)}{\partial x} = 2$ $\sum_{j=1}^{m} (a_j a_j x + a_j b_j y + a_j c_j z - a_j l_j) = 0$, or, in the Gauss notation, ($[aa] = \sum_{j=1}^{m} a_j a_j$ and so forth): $[aa]x + [ab]y + [ac]z = [al]$.

We find the remaining equations analogously. Then the system of normal equations relative to the unknowns x, y, z has the form

$$[aa]x + [ab]y + [ac]z = [al] \quad [ab]x + [bb]y + [bc]z = [bl] \quad [ac]x + [bc]y + [cc]z = [cl].$$
(4.58)

Its solution provides the measurement result – the best values of the unknown quantities using the Legendre criterion:

$$\tilde{x} = \frac{D_a}{D}, \quad \tilde{y} = \frac{D_b}{D}, \quad \tilde{z} = \frac{D_c}{D},$$
(4.59)

where $D = \begin{vmatrix} [aa] & [ab] & [ac] \\ [ab] & [bb] & [bc] \\ [ac] & [bc] & [cc] \end{vmatrix}$ is the determinant of system (4.58),

$D_a = \begin{vmatrix} [al] & [ab] & [ac] \\ [bl] & [bb] & [bc] \\ [cl] & [bc] & [cc] \end{vmatrix}$, and so forth. To estimate the uncertainties of these values, they are substituted into the conditional equations and the residual errors are calculated:

$$v_1 = a_1\tilde{x} + b_1\tilde{y} + c_1\tilde{z} - l_1,$$
$$v_2 = a_2\tilde{x} + b_2\tilde{y} + c_2\tilde{z} - l_2,$$

$$\vdots$$

$$v_m = a_m\tilde{x} + b_m\tilde{y} + c_m\tilde{z} - l_m.$$

The standard uncertainty of the system of conditional equations is equal to

$$u_A\left(\sum_{j=1}^m v_j^2\right) = \sqrt{\frac{\sum_{j=1}^m v_j^2}{m-k}}, \tag{4.60}$$

where m is the number of conditional equations and k is the number of normal equations equal to the number of unknowns. In the case under study $k = 3$.

Thereafter, the type A standard uncertainties of measurement results are determined:

$$u_A(\tilde{x}) = \sqrt{\frac{D_{11}}{D}}u_A\left(\sum_{j=1}^m v_j^2\right), \quad u_A(\tilde{y}) = \sqrt{\frac{D_{22}}{D}}u_A\left(\sum_{j=1}^m v_j^2\right),$$

$$u_A(\tilde{z}) = \sqrt{\frac{D_{33}}{D}}u_A\left(\sum_{j=1}^m v_j^2\right), \tag{4.61}$$

where $D_{11} = \begin{vmatrix}[bb] & [bc] \\ [bc] & [cc]\end{vmatrix}$, $D_{22} = \begin{vmatrix}[aa] & [ac] \\ [ac] & [cc]\end{vmatrix}$, $D_{33} = \begin{vmatrix}[aa] & [ab] \\ [ab] & [bb]\end{vmatrix}$.

After this, using methods studied in Sect. 3.3, the type B standard uncertainties and the extended uncertainties of the measurement results are found. Here the number of degrees of freedom is taken as equal to $m - k$.

As follows from (4.60), the accuracy of measurement results is higher, the greater the number of conditional equations. If this number is small, or differs little from the number of unknowns, then the measurement results are determined with rough approximation. The accuracy of this method also depends significantly on knowledge of the functional dependence of the conditional equations. Their degree of approximation sharply distorts the measurement results. If the conditional equations are rough empirical formulas, then the use of the least squares method will not give good results, even with very precise test data.

One must note that the method presented is widely used in calibrating measuring instruments for experimental determination of calibration dependencies. These are dependencies of the transform function $y(x)$ of the measuring instrument, of a linear form $y = ax + b$, exponential $y = ax^b$, logarithmic $y = a + b\ln x$, and other. In this case, the input quantities are the measurement results of the quantities X and Y at input to and output from the measuring instrument, and the unknown values are the coefficients a and b of the calibration function.

Fig. 11 Angle measurement
on site

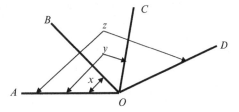

Let us examine the examples of using this method in joint and aggregate measurements.

Example 4.7. Measuring angles at a site (joint measurements)

It is necessary to measure the angles x, y, and z between objects A, B, C, and D at a site (0 is the observer's location point), shown in Fig. 11, and estimate the uncertainty of measurement results, evaluated as type A, with a coverage probability of $P = 0.95$.

Six angles are measured:

$\angle AOB = x = 48.3°$, $\angle AOC = y = 96.8°$, $\angle AOD = z = 152.9°$, $\angle BOC = y - x = 48.5°$, $\angle BOD = z - x = 104.4°$, $\angle COD = z - y = 56.1°$.

The system of conditional equations is as follows:

$$\begin{cases} 1 \times x + 0 \times y + 0 \times z = 48.3, \\ 0 \times x + 1 \times y + 0 \times z = 96.8, \\ 0 \times x + 0 \times y + 1 \times z = 152.9, \\ (-1) \times x + 1 \times y + 0 \times z = 48.5, \\ (-1) \times x + 0 \times y + 1 \times z = 104.4, \\ 0 \times x + (-1) \times y + 1 \times z = 56.1. \end{cases}$$

We find the Gauss coefficients:

$[aa] = 1^2 + 0^2 + 0^2 + (-1)^2 + (-1)^2 + 0^2 = 3,$

$[ab] = 1 \times 0 + 0 \times 1 + 0 \times 0 + (-1) \times 1 + (-1) \times 0 + 0 \times (-1) = -1,$

$[ac] = 1 \times 0 + 0 \times 0 + 0 \times 1 + (-1) \times 0 + (-1) \times 1 + 0 \times 1 = -1,$

$\quad 1 \times x + 0 \times y + 0 \times z =$

$[bb] = 0^2 + 1^2 + 0^2 + 1^2 + 0^2 + (-1)^2 = 3,$

$[bc] = 0 \times 0 + 1 \times 0 + 0 \times 1 + 1 \times 0 + 0 \times 1 + (-1) \times 1 = -1,$

$[cc] = 0^2 + 0^2 + 1^2 + 0^2 + 1^2 + 1^2 = 3,$

$[al] = 1 \times 48.3 + 0 \times 96.8 + 0 \times 152.9 + (-1) \times 48.5$
$\quad + (-1) \times 104.4 + 0 \times 56.1 = -104.6,$

$[bl] = 0 \times 48.3 + 1 \times 96.8 + 0 \times 152.9 + 1 \times 48.5 + 0 \times 104.4 + (-1) \times 56.1 = 89.2,$

$[cl] = 0 \times 48.3 + 0 \times 96.8 + 1 \times 152.9 + 0 \times 48.5 + 1 \times 104.4 + 1 \times 56.1 = 313.4.$

Hence, the system of normal equations (4.58) takes the form:

$$\begin{cases} 3x - y - z = -104.6, \\ -x + 3y - z = 89.2, \\ -x - y + 3z = 313.4. \end{cases}$$

Table 17 Measurements for calibrating a gas analyzer

Index i of measurement	Volume share C, ppm	Output signal of instrument W, V
1	0.5	0.5
2	1.0	3.2
3	1.5	5.9
4	2.0	8.5

The determinants of this system are: $D = 16$, $D_a = 773.6$, $D_b = 1549$, $D_c = 2446$. We find the best measurement results:

$$\tilde{x} = \frac{773.6}{16} = 48.350°, \quad \tilde{y} = \frac{1546}{16} = 96.813°, \quad \tilde{z} = \frac{2446}{16} = 152.875°.$$

Now we calculate the residual errors:

$$v_1 = 48.35 - 48.3 = 0.05°,$$
$$v_2 = 96.813 - 96.8 = 0.013°,$$
$$v_3 = 152.875 - 152.9 = -0.025°,$$
$$v_4 = -48.35 + 96.813 - 48.5 = -0.037°,$$
$$v_5 = -48.35 + 152.875 - 104.4 = 0.125°,$$
$$v_6 = -96.813 + 152.875 - 56.1 = -0.038°.$$

The standard uncertainty of the system of conditional equations is equal to

$$u_A \left(\sum_{j=1}^{6} v_j^2 \right) = \sqrt{\frac{\sum_{j=1}^{6} v_j^2}{6 - 3}} = \sqrt{\frac{0.022}{3}} = 0.086°.$$

Further, in formulas (4.61) $D_{11} = D_{22} = D_{33} = \begin{vmatrix} 3 & -1 \\ -1 & 3 \end{vmatrix} = 8$, and the type A standard uncertainties of the measurements results will be as follows:

$$u_A(\tilde{x}) = u_A(\tilde{y}) = u_A(\tilde{z}) = \sqrt{\frac{8}{16}} \times 0.086 = 0.061°.$$

In accordance with Table 5 the coverage factor, for $P = 0.95$ and number of degrees of freedom $n = 3$, is equal to $k = 3.31$. Hence the extended uncertainty of the results of measuring the angles is $U_A(\tilde{x}) = U_A(\tilde{y}) = U_A(\tilde{z}) = 3.31 \cdot 0.061 = 0.202°$.

Hence, with a coverage probability of 0.95, the measured angles are within the following bounds: $x = (48.350 \pm 0.2)°$, $y = (96.813 \pm 0.2)°$, $z = (152.875 \pm 0.2)°$.

Example 4.8. Calibration of a gas analyzer (joint measurements)

Table 17 shows measurement results produced in calibrating a gas analyzer. Find the linear calibration characteristics of the instrument $W = x + Cy$ and its extended uncertainty at a coverage probability of $P = 0.95$. The type B relative extended uncertainty, due to uncertainty in the attributed values of unit prototype, is $U_{B.rel}(C) = 0.05$ for $P = 0.95$. Estimate also the uncertainty of the calibration characteristics in operation. Instability of the additive and of the multiplicative errors of the gas analyzer are estimated by the extended uncertainties $U_B(x) = 0.1$ V and $U_B(y) = 0.05$ V/ppm for $P = 0.95$.

The system of conditional equations will be as follows:

$$\begin{cases} 1x + 0.5y = 0.5, \\ 1x + 1.0y = 3.2, \\ 1x + 1.5y = 5.9, \\ 1x + 2.0y = 8.5. \end{cases}$$

Gauss coefficients: $[aa] = 1^2 + 1^2 + 1^2 + 1^2 = 4$,

$[ab] = 1 \times 0.5 + 1 \times 1.0 + 1 \times 1.5 + 1 \times 2.0 = 5.0$,

$[bb] = 0.5^2 + 1.0^2 + 1.5^2 + 2.0^2 = 7.5$,

$[al] = 1 \times 0.5 + 1 \times 3.2 + 1 \times 5.9 + 1 \times 8.5 = 18.1$,

$[bl] = 0.5 \times 0.5 + 1.0 \times 3.2 + 1.5 \times 5.9 + 2.0 \times 8.5 = 29.30$.

The system of normal equations has the form:

$$\begin{cases} 4x + 5y = 18.1, \\ 5x + 7.5y = 29.30. \end{cases}$$

Determinants of this system: $D = 5.0$, $D_a = -10.75$, $D_b = 26.70$. The best measurement results are: $\tilde{x} = -10.75/5 = -2.15$ V, $\tilde{y} = 26.70/5 = 5.34$ V/ppm. Now we calculate the residuals (residual errors):

$$v_1 = -2.15 + 0.55 \times 5.34 - 0.5 = 0.02 \text{ V},$$
$$v_2 = -2.15 + 1.0 \times 5.34 - 3.2 = -0.01 \text{ V},$$
$$v_3 = -2.15 + 1.5 \times 5.34 - 5.9 = -0.04 \text{ V},$$
$$v_4 = -2.15 + 2.0 \times 5.34 - 8.5 = 0.03 \text{ V}.$$

The standard uncertainty of the system of conditional equations is equal to

$$u_A \left(\sum_{j=1}^{4} v_j^2 \right) = \sqrt{\frac{\sum_{j=1}^{4} v_j^2}{4 - 2}} = \sqrt{\frac{0.003}{2}} = 0.039 \text{ V}.$$

Further, in formulas (4.61) $D_{11} = 7.5$, $D_{22} = 4$, and the type A standard uncertainties of the measurement results will be as follows: $u_A(\tilde{x}) = \sqrt{7.5/5} \times 0.039 = 0.048$ V, $u_A(\tilde{y}) = \sqrt{4/5} \times 0.039 = 0.035$ V/ppm.

Fig. 12 Calibration characteristic of a gas analyzer and its bounds, corresponding to a coverage probability of 0.95

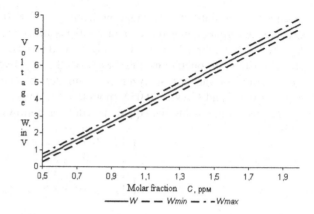

In accordance with Table 5, the coverage coefficient for $P = 0.95$ and number of degrees of freedom $n = 2$ equals $k = 4.53$.

Let us find the estimates of the extended uncertainty of the calibration results. By the conditions of the task, the uncertainty at the beginning point of the scale is caused only by type A uncertainty. Hence $U(x) = U_A(x) = ku_A(x) = 4.53 \times 0.048 = 0.22$ V. Uncertainty of the calibration coefficient is caused by type A uncertainty, estimated as at least $(u_A(\bar{y}) = 0.035$ V/ppm$)$, and by uncertainty of the attributed values of the prototype. The standard uncertainty caused by the latter source is estimated, with equal probability distribution and $P = 0.95$, with a value $u_B(y) = \frac{0.05}{1.65} = 0.03$ V/ppm. The total standard uncertainty is equal to

$$u(y) = \sqrt{u_A^2(\bar{y}) + u_B^2(y)} = \sqrt{0.035^2 + 0.03^2} = 0.046 \text{ V/ppm}.$$

The effective number of degrees of freedom by formula (3.18) equates to

$$\nu_{eff} = \frac{0.046^4}{0.035^4/2 + 0.03^4/\infty} = 6.0.$$

In accordance with Table 5, this value corresponds to a coverage factor of $k = 2.52$. With this, $U(y) = 2.52 \times 0.046 = 0.116$ V/ppm. Since the distributions of the uncertainty of values of x and y are mutually independent, one may estimate the extended uncertainty of the calibration characteristics at point C of the range of measurements by their mean square sum: $U_{cal}[W(C)] = \sqrt{0.22^2 + (0.116C)^2}$. Figure 12 shows the graphs of the calibration function $W(C) = - - = (-2.15 + 5.34C)$ V, and their bounds $W_{min}(C) = W(C) - U[W(C)]$, $W_{max}(C) = W(C) + U[W(C)]$, corresponding to a coverage probability of 0.95.

Now let us estimate the instability of the calibration characteristic. The standard uncertainties due to the instability of the additive and of the multiplicative errors of the gas analyzer are equal to $u_B(x) = 0.1/1.65 = 0.06$ V and

$$\tilde{x} = \frac{n \sum_{i=1}^n a_i l_i - \sum_{i=1}^n a_i \sum_{i=1}^n l_i}{n \sum_{i=1}^n a_i^2 - \left(\sum_{i=1}^n a_i\right)^2},$$

$$\tilde{y} = \frac{1}{b} \frac{\sum_{i=1}^n a_i^2 \sum_{i=1}^n l_i - \sum_{i=1}^n a_i \sum_{i=1}^n a_i l_i}{n \sum_{i=1}^n a_i^2 - \left(\sum_{i=1}^n a_i\right)^2} \text{ V/ppm}.$$

The total standard uncertainties of these values are equal to $u(x) = \sqrt{0.048^2 + 0.06^2} = 0.077$ V and $u(y) = \sqrt{0.046^2 + 0.03^2} = 0.055$ V/ppm. The effective number of their degrees of freedom are, by formula (3.18), equal to

$$v_{\text{eff}}(x) = \frac{0.077^4}{0.048^4/2 + 0.06^4/\infty} = 13.2$$

and

$$v_{\text{eff}}(y) = \frac{0.055^4}{0.035^4/2 + 2 \times 0.03^4/\infty} = 12.2.$$

In accordance with Table 5, these correspond to coverage factors of $k(x) = 2.23$ and $k(y) = 2.25$. With this, $U(x) = 2.23 \times 0.077 = 0.172$ V and $U(y) = 2.25 \times 0.055 = 0.124$ V/ppm. The extended uncertainty of the calibration characteristic at point C of the range of measurements equates to $U_{\text{oper.}}[W(C)] = \sqrt{0.172^2 + (0.124C)^2}$.

Example 4.9. Environmental appraisal of occupational noise (joint measurements)[2]

Occupational noise is the name given to undesirable and irregular set of sounds that impact negatively on the worker's state of health. Sounds differing in frequency and intensity spread out as longitudinal vibrations in the air. The human ear hears sounds in the range of frequencies from 16–20 Hz to 20,000 Hz, reacting not to the absolute value of sound pressure, but to its relative change. And since there exists an approximately logarithmic relationship between the energy of excitation and sound perception, for convenience a value has been introduced for sound measurement – the level L of the sound intensity in dB, which is defined by the formula

$$L = 10 \times \lg_{10}\left(\frac{I}{I_0}\right), \tag{4.62}$$

---

[2] Laboratory work of the Department of Information Systems for Environmental Safety, SPbGPU.

Table 18 Test record sheet

Index i of measurement	Distance r_i, m	ai, $1/m^2$	Level L_i, dB	Sound intensity I_i, W/m^2	Residual v_i, W/m^2
1	2	3	4	5	6
1	1	0.0796	111.5	0.01412	−0.00628
2	2	0.0199	108.0	0.010631	−0.00081
3	3	0.0088	105.0	0.00316	−0.00042
4	4	0.0050	102.0	0.00158	0.00020
5	5	0.0032	98.0	0.00063	0.00070
6	6	0.0022	91.0	0.00013	0.00096
7	7	0.0016	87.0	0.00005	0.00089
8	8	0.0012	83.0	0.00002	0.00083

where I is the sound intensity in W/m^2 and $I_0 = 1 \times 10^{-12}$W/m^2 is the threshold sound intensity.

The presence of a great number of reflecting surfaces in the room substantially increases the level of sound intensity. The sound intensity in such a room is estimated by the formula

$$I = N\left[\frac{1}{4\pi r^2} + \frac{4(1 - \alpha_{mean})}{\alpha_{mean}S}\right], \tag{4.63}$$

where N is the power of the sound source in Watts, r is the distance to the sound source in meters, S is the total area of enclosing surfaces in m^2, and α_{mean} is the mean coefficient of source absorption in the room.

In the process of studying occupational noise in a room, it is possible to measure S, the distance to the source of noise at r_i different points of the room, and the sound intensity I_i at these points. From this data, the power N of the source of noise is measured, along with the public health characteristic of the room – its mean factor α_{mean} of sound absorption.

From the metrological point of view, these are joint measurements. However, the dependence between the unknown quantities and the measurand is not linear. Hence we transform it by the method of replacing variables. Let us introduce the following notation:

$$a = \frac{1}{4\pi r^2}, \; b = \frac{4}{S}, \; x = N, \; y = N \times \frac{1 - \alpha_{mean}}{\alpha_{mean}}, \; l = I = I_0 \times 10^{L/10}.$$

Above all, it is essential to measure the constant quantity b. The result of measuring the area of enclosing surfaces is $S = (90.5 + 0.5)M^2$ (the uncertainty is distributed by the equal probability law). Hence $b = 4/S = 0.0442$ M^{-2}. Then points i are selected in the room, at which measurements are taken of the distances r_i to the sound source, and with the aid of a sound meter, the level of sound intensity L_i. Results of these measurements are shown in columns 2 and 4 of Table 18. From the values r_i, L_i, the quantities a_i, l_i are calculated and shown in columns 3 and 5 of this table.

The system of conditional equations in this case has the form:

$$\begin{cases} a_1x + by = l_1, \\ a_2x + by = l_2, \\ \vdots \\ a_8x + by = l_8. \end{cases}$$

Using the method described above, this system is transformed in a system of normal equations

$$\begin{cases} [aa]x + [ab]y = [al], \\ [ab]x + [bb]y = [bl], \end{cases}$$

where

$$[aa] = \sum_{i=1}^{n} a_i^2, \quad [ab] = \sum_{i=1}^{n} a_ib = b\sum_{i=1}^{n} a_i, \quad [bb] = \sum_{i=1}^{n} b^2 = nb^2,$$

$$[al] = \sum_{i=1}^{n} a_il_i, \quad [bl] = \sum_{i=1}^{n} bl_i = b\sum_{i=1}^{n} l_i.$$

The solution of this system of equations is $\tilde{x} = D_a/D$, $\tilde{y} = D_b/D$, where the main determinant is

$$D = \begin{vmatrix} [aa] & [ab] \\ [ab] & [bb] \end{vmatrix} = b^2\left[n\sum_{i=1}^{n} a_i^2 - \left(\sum_{i=1}^{n} a_i\right)^2\right],$$

$$D_a = \begin{vmatrix} [al] & [ab] \\ [bl] & [bb] \end{vmatrix} = b^2\left[n\sum_{i=1}^{n} a_il_i - \sum_{i=1}^{n} a_i\sum_{i=1}^{n} l_i\right],$$

$$D_b = \begin{vmatrix} [aa] & [al] \\ [ab] & [bl] \end{vmatrix} = b\left[\sum_{i=1}^{n} a_i^2\sum_{i=1}^{n} l_i - \sum_{i=1}^{n} a_i\sum_{i=1}^{n} a_il_i\right].$$

Hence,

$$\tilde{x} = \frac{n\sum_{i=1}^{n} a_il_i - \sum_{i=1}^{n} a_i\sum_{i=1}^{n} l_i}{n\sum_{i=1}^{n} a_i^2 - \left(\sum_{i=1}^{n} a_i\right)^2},$$

$$\tilde{y} = \frac{1}{b}\frac{\sum_{i=1}^{n} a_i^2\sum_{i=1}^{n} l_i - \sum_{i=1}^{n} a_i\sum_{i=1}^{n} a_il_i}{n\sum_{i=1}^{n} a_i^2 - \left(\sum_{i=1}^{n} a_i\right)^2}. \tag{4.64}$$

Substituting into these equations $b = 0.0442$ m^{-2}, $n = 8$, $\sum_{i=1}^{8} a_i = 0.1216 \times$ 1/m^2, $\sum_{i=1}^{8} a_i^2 = 0.00686 \times 1/$m^4, $\sum_{i=1}^{8} l_i = 0.026$ W/m^2, $\sum_{i=1}^{8} a_i l_i = 0.00129$ W/m^4, we derive $\tilde{x} = 0.25$ W, and $\tilde{y} = 0.012$ W.

For estimation of the uncertainty of these values, the residuals of the conditional equations, as presented in column 6 of Table 18, are calculated. The standard uncertainty of the system of conditional equations is, in accordance with (4.60), equal to

$$ u_A \left(\sum_{i=1}^{8} v_i^2 \right) = \sqrt{\frac{\sum_{i=1}^{8} v_i^2}{8-2}} = \sqrt{\frac{4.3 \times 10^{-5}}{6}} = 0.00268 \text{ W/m}^2. $$

$$ D_{11} = [bb] = nb^2, \quad D_{22} = [aa] = \sum_{i=1}^{n} a_i^2, $$

$$ u_A(\tilde{x}) = \sqrt{\frac{D_{11}}{D}} \times u_A \left(\sum_{i=1}^{8} v_i^2 \right) $$

$$ = \sqrt{\frac{8}{8 \sum_{i=1}^{8} a_i^2 - \left(\sum_{i=1}^{8} a_i \right)^2}} u_A \left(\sum_{i=1}^{8} v_i^2 \right) $$

$$ = \sqrt{\frac{8}{8 \times 0.00686 - 0.1216^2}} \times 0.00268 $$

$$ = 0.038 \text{ W} $$

and

$$ u_A(\tilde{y}) = \sqrt{\frac{D_{22}}{D}} \times u_A \left(\sum_{i=1}^{n} v_i^2 \right) = \frac{1}{b} \sqrt{\frac{\left[\sum_{i=1}^{n} a_i^2 \right]}{n \sum_{i=1}^{n} a_i^2 - \left(\sum_{i=1}^{n} a_i \right)^2}} \times u_A \left(\sum_{i=1}^{n} v_i^2 \right) $$

$$ = \frac{1}{0.0442} \times \sqrt{\frac{0.00686}{8 \times 0.00686 - 0.1216^2}} \times 0.00268 = 0.025 \text{ W}. $$

Let us now turn to the measurand. Making an inverse replacement of variables, we obtain the values of the measurand: $\tilde{N} = \tilde{x} = 0.25$ W, and $\tilde{y}\tilde{\alpha}_{mean} = \tilde{x}(1 - \tilde{\alpha}_{mean})$, from which $\tilde{\alpha}_{mean} = \frac{\tilde{x}}{\tilde{x}+\tilde{y}} = \frac{0.25}{0.25+0.012} = 0.95$. We calculate the standard uncertainties of the values $\tilde{N}$ and $\tilde{\alpha}_{mean}$. $u_A(\tilde{N}) = u_A(\tilde{x}) = 0.038$ W. To estimate the uncertainty $\tilde{\alpha}_{mean}$ we expand the function $\alpha_{mean} = f(x, y) = \frac{x}{x+y}$ into a Taylor series about the value $\tilde{\alpha}_{mean} = \frac{\tilde{x}}{\tilde{x}+\tilde{y}}$, limiting it just to its linear part. Then, considering that

$$ \Delta f(x, y) = f(\tilde{x}, \tilde{y}) - f(x, y) = \frac{\partial f(x, y)}{\partial x} \bigg|_{x=\tilde{x}, \, y=\tilde{y}} \Delta x + \frac{\partial f(x, y)}{\partial y} \bigg|_{x=\tilde{x}, \, y=\tilde{y}} \Delta y, $$

where

$$\frac{\partial f(x,y)}{\partial x} = \frac{y}{(x+y)^2},$$

and

$$\frac{\partial f(x,y)}{\partial y} = -\frac{x}{(x+y)^2},$$

as well as the fact that the quantities x and y are not correlated, we derive:

$$u_A(\tilde{\alpha}_{mean}) = \frac{\sqrt{\tilde{y}^2 u_A^2(\tilde{x}) + \tilde{x}^2 u_A^2(\tilde{y})}}{(\tilde{x}+\tilde{y})^2}$$

$$= \frac{\sqrt{0.0442^2 \times 0.038^2 + 0.25^2 \times 0.025^2}}{(0.0442+0.25)^2} = 0.075.$$

The standard uncertainty of the result of measuring the area of the enclosing surfaces is $u_B(S) = \frac{0.5}{1.73} = 0.3\,\text{m}^2$. Consequently, the standard uncertainty of the quantity b is equal to $u_B(b) = \left|\frac{\partial b}{\partial S}\right| \times u_B(S) = b \times \frac{u_B(S)}{S} = 0.0442 \times \frac{0.3}{90} \cong 0.00015\,\text{m}^{-2}$. Formulas (4.64) show that this affects only the measurement results α_{cp}. $u_B(\tilde{\alpha}_{mean}) = \left|\frac{\partial \alpha_{mean}}{\partial y}\right| \times \left|\frac{\partial y}{\partial b}\right| \times u_B(b) = \frac{\tilde{x}}{(\tilde{x}+\tilde{y})^2} \times \frac{\tilde{y}}{b} \times u_B(b) = \frac{0.25 \times 0.012}{(0.25+0.12)^2 \times 0.0442} \times 0.00015 \cong 0.00015$ is insignificant by comparison with $u_A(\tilde{\alpha}_{mean}) = 0.075$.

Hence, $u(\tilde{N}) = u_A(\tilde{N}) = 0.038\,\text{W}$, and $u(\tilde{\alpha}_{mean}) \cong u_A(\tilde{\alpha}_{mean}) = 0.075$.

The extended uncertainty is determined by formulas (3.16) and (3.17), in which $v_{eff}(\tilde{N}) = n - 2 = 6$, and, by formula (3.18),

$$v_{eff}(\tilde{\alpha}_{mean}) = (n-2)\frac{(\tilde{x}+\tilde{y})^8 u^4(\tilde{\alpha}_{mean})}{\tilde{y}^4 u_A^4(\tilde{x}) + \tilde{x}^4 u_A^4(\tilde{y})}$$

$$= 6 \times \frac{0.262^8 \times 0.075^4}{0.012^4 \times 0.038^4 + 0.25^4 \times 0.025^4} = 6 \times 0.46 = 2.76.$$

In accordance with Table 5, coverage factors of $k(\tilde{N}) = 2.52$, $k(\tilde{\alpha}_{mean}) = 3.0$ correspond to these values. Hence $U(\tilde{N}) = 2.52 \times 0.038 = 0.1\,\text{W}$ and $U(\tilde{\alpha}_{mean}) = 3.0 \times 0.075 = 0.225$.

So, the intensity of the noise source is $N = (0.25 \pm 0.1)\,\text{W}$, the expected value of the mean coefficient of sound absorption of the room is $\tilde{\alpha}_{mean} = 0.95$, and the extended uncertainty of this quantity is within the limits $\alpha_{mean} = [0.725 - 1]$. The estimates presented for extended uncertainty were derived under the assumption of their correspondence to Student's distribution and a coverage probability of $P = 0.95$.

Chapter 5
The International System of Units: SI

5.1 Systems of Units and the Principles of Their Formation

Unity of measurements is understood as consistency on the dimensions of the units of all quantities. This is evident when recalling the possibility of measuring one and the same quantity by direct and indirect measurements. Such consistency is achieved by the creation of a system of units. But, although the advantages of a system of units were recognized long ago, the first system of units appeared only at the end of the eighteenth century. This was the celebrated metric system (meter, kilogram, second), authorized March 26, 1791 by the Constituent Assembly of France. The first scientifically based system of units, as a set of arbitrary base units and dependent units derived from them, was presented in 1832 by Carl Gauss. He constructed a system of units, called absolute units, at the base of which were three arbitrary and mutually independent units: millimeter, milligram, and second. A further development of Gauss' system was the CGS system (centimeter, gram, second) that appeared in 1881 and was convenient for use in electromagnetic measurements, as well as various modifications to it.

Development of industry and trade in the era of the first industrial revolution required unification of units on an international scale. The beginning of this process was set May 20, 1875 with the signing by 17 countries (including Russia, Germany, the USA, France, and England) of the Metric Convention, which many other countries joined later. International cooperation in the area of metrology was established in accordance with this Convention. The International Bureau of Weights and Measures (BIPM) was created at Sèvres, a suburb of Paris, for the purpose of conducting international metrological research and storing international standard samples. The International Committee for Weights and Measures was founded to direct the BIPM, and includes consultative committees on the units and series of forms of measurements. To resolve the most important issues of international metrological cooperation, international conferences called General Conferences on Weights and Measures (CGPM) began to be conducted regularly. All countries that subscribed to the Metric Convention received prototypes of the international standard samples of

A.E. Fridman, *The Quality of Measurements: A Metrological Reference*,
DOI 10.1007/978-1-4614-1478-0_5, © Springer Science+Business Media, LLC 2012

length (the meter) and mass (the kilogram). Periodic comparisons of these national standard samples with the international standard samples stored at BIPM were also organized. In this way the metrical system of units first obtained international recognition. However, after the signing of the Metric Convention, systems of units for other realms of measurements were developed – CGS, CGS electrostatic, CGS electromagnetic, meter-ton-second, meter-kilogram-second, meter-kilogram-force-second. Again the issue of unity of measurements arose, this time between the different realms of measurement. Both in 1954 at the 10th CGPM in preliminary form, and in 1961 at the 11th CGPM conclusively, the International System of Units (SI) was adopted, and has been operative with insignificant modifications until the present. Several times modifications and supplements have been introduced into it at succeeding sessions of the CGPM. Currently, the SI System of Units is regulated by ISO Standard 31 [27] and in essence is an international regulation mandatory for use. In Russia, ISO Standard 31 was confirmed as State Standard GOST 8.417–02 [28].

The SI System of Units was formed in accordance with the general principle of forming a system of units as proposed by Carl Gauss in 1832. In accordance with this, all physical quantities are divided into two groups: quantities accepted as independent from other quantities, which are called base quantities; all other quantities, called derived, which are expressed through the base quantities and already-defined derived quantities using physical equations. From this follows also the classification of units: the units of base quantities are the based units of the system, and the units of derived quantities are derived units.

And so, there is first formulated *a system of quantities – a set of quantities that is formulated in accordance with the principle wherein some quantities are accepted as independent and others are functions of the independent quantities A quantity entering the system of quantities and conventionally taken as independent of other quantities of this system is called a base quantity. A quantity entering the system of quantities and defined through base quantities and already-defined derived quantities is called a derived quantity.*

The unit of a base quantity is called a base unit. A derived unit is the unit of a derived quantity of the given system of units, formulated in accordance with an equation linking it with the base units, or with the base units and already-defined derived units.

In this manner is formed the system of units of quantities – *a set of base and derived units of the specified system of quantities.*

5.2   Base Units of the SI

The minimum number of base units is determined as follows. Let there be, between the numerical values of N diverse physical quantities, n coupling equations into which known physical laws enter. Each equation has its own coefficient of proportionality to which one may assign any value, and in particular may equate it to unity. Hence, these coefficients are known numbers, but the quantities are unknowns.

The number of quantities is always greater than the number of coupling equations: $N > n$. In order to find the values of these N quantities from the existing system of n equations, it is necessary to add to this system another $N - n$ equations, i.e., add the values of $N - n$ quantities. Consequently, these values must be units independent of the other quantities.

Hence the minimum number of base quantities is equal to $N - n$. Such a system of quantities is called theoretically optimal. In practice, the number of base quantities is greater: $N - n + p$. In this case, the number of unknowns in a system consisting of n equations is less than n, since it is equal to $N - (N - n + p) = n - p$. Consequently, this system of equations is over-determined and has an infinite number of solutions. In order to ensure a unique solution, it is necessary to add p equations. Hence once must take p coefficients as equal not to one, but to some numbers determined as a result of solving this system of equations. These coefficients are called the fundamental physical constants. Examples of such constants are the gravitational constant, and the Planck and Boltzmann constants.

Which units should be chosen as the base units of the system, out of the total set? There is no unambiguous answer to this question. Resolution of this issue is guided by practical expediency, taking into account the various arguments such as:

- the selection as base units of the minimum number of quantities that reflect the most common properties of matter,
- the clearest reflection of the mutual interaction of the base quantities that belong to different realms of physics,
- the high accuracy of the reproduction of units and the transfer of their dimensions,
- the ease of formation of derived quantities and units,
- the continuity of the units; i.e., the preservation of their dimensions and designations when introducing modifications into the system of units.

These criteria often come into contradiction with each other. Hence the selection of a decisive version is the subject of agreement strengthened by decision of the CGPM.

Let us examine how this has been implemented in the SI system. In measurements of space and time, the foundation is the equation of motion $v = K(dl/dt)$, where v is velocity, l is distance, t is time, and K is an arbitrarily selected coefficient whose value depends on the selection of units. Addition of other quantities (area, volume, acceleration, etc.) is accompanied by the addition of a corresponding coupling equation, and hence changes basically nothing. So $N = 3$ and $n = 1$, and consequently $N - n = 2$. Hence two base quantities are accepted: distance l and time t and, correspondingly, two base units: the meter and the second. In 1983, the speed of light in vacuum, the true value of which is in principle unknown, was assigned an exact value by agreement: 299,792,458 m/s. Here *the unit of distance – the meter –* was defined as *the distance through which light passes in vacuum in 1/299,792,458 of a second*. With this definition of a meter, the coefficient of proportionality K in the equation of motion remained equal to 1. If the old definition of a meter had been preserved, and at the same time the constancy of the speed of light had been postulated, it would not have been possible to take K as equal to 1, and it would have been a fundamental physical constant.

In accordance with its theoretical definition, the meter is reproduced by measuring the wavelength of the ultra-stable helium–neon and argon lasers. The relative uncertainty of its reproduction does not exceed 1×10^{-9}. Another base unit – the *second* – is defined as the time equal to 9,192,631,770 *periods of the radiation corresponding to the transition between the two hyperfine levels of the ground state of the Cesium-133 atom*. It is reproduced by measuring the frequency of radiation of Cesium-133 atoms with a relative uncertainty not exceeding 1×10^{-13}. It is evident that the uncertainty in measuring the meter is four orders of magnitude higher than the uncertainty in defining the second. Comparison of the standards for distance and frequency is theoretically possible, based on the well-known relationship $L = h/v$ for wavelength L and frequency v (where h is the Planck constant). This will be feasible in practice when the leading metrological centers of the world are equipped with advanced comparators suitable for comparing samples of frequencies at the junction of the optical range and the radio frequency range (the so-called ROFBs, or radio-frequency optical bridges). Hence it is completely likely that the prospects are that, with the goal of refining the accuracy of measurements of distance, its unit will be rendered by combining it with the standards for frequency. In that event, distance will in fact be a derived quantity, dependent on time. But, most likely, the meter will formally remain, as before, a base unit in the SI.

In the transition to mechanical measurements, the equation of motion is fulfilled by Newton's Law $F = K_1 ma$ (F is force, m is mass, and a is acceleration) and the law of universal gravitation $F = K_2(m_1 m_2)/r^2$ (r is the distance between the bodies). If we set $K_1 = K_2 = 1$, then two quantities (F and m) and two coupling equations are added. Consequently, it is theoretically possible not to introduce additional independent quantities, but consider the mass as a derived quantity. For example, with $m = m_1 = m_2$ it follows from these equations that $m = ar^2$. If we render this equation, the unit of mass will be the mass of a material point that transmits a unit acceleration to another material point located at unit distance. However, the accuracy of reproducing such a unit would be very low. Hence a third base unit is introduced – *the unit of mass (the kilogram), which is the mass of the international prototype kilogram located at the BIPM*. Here, in order to preserve the uniqueness of the solution of the system of quantities, it was necessary to introduce into the law of universal gravitation a coefficient of proportionality that differs from unity: the gravitational constant $\gamma = (6.670 \pm 0.041) \times 10^{-11}$ (N $\times$ m^2)/kg^2. γ is a fundamental physical constant.

The unit of mass has remained until the present the only base unit that is reproduced not on the basis of a physical effect but by an artificially created measure (a material prototype). B.N. Taylor (NIST) already noted in 1977 five basic problems with this definition of the unit: there is the possibility of damage or even loss; the prototype accumulates impurities and cannot easily be cleaned of them; the prototype deteriorates in an unknown manner; it is not possible to use it often because of the fear of wearing it down; and it is accessible only in one laboratory.

In addition, to the accuracy of the prototype (relative uncertainty to the degree of 1×10^{-7}) are already attached the requirements for accuracy of a number of measurement techniques that use the measurement of mass. Hence at the current time, work is proceeding to search for an optimal method of over-defining the unit of mass on the basis of some physical phenomenon. In 2005 the CGPM adopted Recommendation № 1 (CI–2005) "Preparative steps towards new definitions of the kilogram, the Ampere, the Kelvin and the mole in terms of fundamental constants". [29]. Hence the day is not far off when mass will be reproducible through other quantities of the SI, formally, openly, and remaining a base quantity.

In thermodynamics, four coupling equations define the same quantity, the thermodynamic temperature T:

- the Mendeleev-Clapeyron law $pv = \frac{m}{M}RT$ (p is pressure, M is molar weight, and R is the universal gas constant);
- the equation of the average kinetic energy of the forward motion of a molecule of ideal gas $W = 1.5k_BT$ (k_B is the Boltzmann constant);
- the Stefan-Boltzmann law $W_R = \sigma T^4$ (σ is the Stefan-Boltzmann constant), linking the volumetric density of electromagnetic radiation W_R with temperature;
- Wien's displacement law $\lambda_m = b/T$ (b is the Wien constant), linking wavelength of the maximum radiation λ_m with temperature.

In doing so, one quantity (temperature) and four equation coefficients are added. From the theoretical point of view of these equations, it is sufficient to define temperature as a derived quantity. However, the historical development of science and the exceptional position held by temperature have made it expedient to accept it as a base quantity. The selection of temperature as among the base quantities resulted in the regulation of a new fundamental physical constant – the Boltzmann constant k_B and the definition of the other constants of this area of measurements (R, σ, and b), using it and other constants.

The unit of thermodynamic temperature – the *kelvin (K)* – is 1/273.16 *part of the thermodynamic temperature of the triple point of water*. Measuring temperatures in the thermodynamic scale is complex and laborious and requires unique and expensive equipment. In addition, random measurement errors significantly increase the random errors of the platinum resistance thermometers. Hence practical temperature scales are used in measurements, made by interpolating the greater number of reference points including triple points, melting points and solidifying points of various pure substances. At the present time, the ITS-90 International Temperature Scale, adopted at the 17th Session of the Consultative Committee for Thermometry in 1989, is operative as the closest possible to the thermodynamic scale. In accordance with [29], the definition of the kelvin will likewise be refined.

In electromagnetism, two equations must be added to the equations of mechanics: Coulomb's law $F = K(q_1q_2/r^2)$ (q is electrical charge) and the mutual interaction of current I with electrical charge $I = q/t$. In these two equations, three new quantities q, I, and K are introduced. All other units of electrical quantities are defined by the laws of electrostatics and electrodynamics. Consequently, $N - n = 1$. Hence in the SI system a new base unit is introduced: the unit of current, the ampere (A). Then the charge is expressed by the relationship $q = It$. The coefficient of proportionality in Coulomb's law takes the form $K = 1/4\pi\varepsilon_0$, where ε_0 is a fundamental physical constant that has received the name "electrical permittivity of vacuum". With this, Coulomb's law is expressed as $F = q_1 q_2/4\pi\varepsilon_0 r^2$.

An ampere is defined as the force of an unchanging current which, in passing through two parallel wires of infinite length and vanishingly small circular cross-sectional area placed in a vacuum 1 m apart, would cause on each 1-meter segment of each conductor an electrodynamic force of 2×10^{-7} N. Reproduction of this unit based on the given definition is possible using specialized measurement instruments called Ampere balances. However, the accuracy of these instruments that has been achieved to date is not very high: the relative standard uncertainty is at least 4×10^{-6}. In that regard, the ampere is reproduced using standard samples of volt and ohm with relative standard uncertainty no greater than 1×10^{-8}. Hence at present there is a contradictory situation in electrical measurements: theoretically the ampere is considered a base unit of the SI, and the volt is a derived unit. In practice, the volt is reproduced based on the Josephson effect independently of other units. The ampere, on the other hand, depends on the volt and in fact is a derived unit. Consequently, the units of electrical quantities used currently are outside the SI. In accordance with [29], this situation will be corrected in the next few years.

There is an unambiguous mutual link between energy and luminous quantities and, strictly speaking, no base unit needs to be introduced to measure luminous quantities. However, considering the historically existing number of base units, a decision was rendered to introduce a unit of the power of light – the candela (cd). In accordance with the now operative definition, *the candela is the power of light emitted in a specified direction of the source, of monochromatic radiation at* 540×10^{12} *Hz, the luminous power of which is 1/683 W/sr in this direction.*

The mole was the last among the base units of the SI to be included in 1971, by decision of the 14th CGPM. *The unit of amount of a substance – the mole – is the amount of the substance in a system containing as many structural elements as there are in 0.012 kg of Carbon-12. Structural elements can be atoms, molecules, ions, electrons and other particles, and other groups of particles.* Standard samples of a mole have never been created and never will be created, since the physical quantity which it represents (the number of structural components) is in its essence a countable quantity. Hence the mole also is a countable quantity, equal to the Avogadro number, $6.02214199(47) \times 10^{23}$ particles. The mole is referenced to the number of base units, in connection with the fact that it helps form the most important derived units of physical chemistry: molar concentration and molar fraction of a component. However, these derived units cannot be reproduced with measuring instruments that preserve moles, since there simply do not exist such instruments. They are reproduced by relying on other base units – the meter and kilogram – by the gravimetric method using the known values of the molar weights of the components. It is not possible to explain its inclusion into the list of SI base units by any other reason

Table 19 Base SI units

Quantity			Unit	
Designation	Dimension	Recommended symbol	Designation	Symbol
Length	L	l	Meter	m
Mass	M	m	Kilogram	kg
Time	T	t	Second	s
Power of electrical current	T	I	Ampere	A
Thermodynamic temperature	θ	T	Kelvin	K
Light power	J	J	Candela	cd
Amount of substance	N	n	Mole	mol

than convenience in using the mole in chemical measurement tasks [6]. Hence the advisability of this decision remains until now under discussion.

And so the units listed in Table 19 are taken as the base units in the SI.

The discussion presented above has shown that the basic principle for constructing a system of units is its convenience of use. It is predominantly such concepts that serve as a guide in determining the number of base units, selecting base units, and selecting their dimensions. All other principles are turned away from being used as base principles. Predominantly for this reason, in many cases the habituation of specialists and the population to a traditional system of quantities and units is a stronger argument than is the possibility of selecting as a base unit one that is reproduced with the highest accuracy. This is also underlined by the fact that in a number of countries a system of units differing from the SI is used. So for example, English-speaking countries tenaciously continue to use their own archaic systems of units (inch, yard, foot, mile, pound, degrees Fahrenheit, etc.), which are banned by the ISO 31 standard. There are also more valid cases – the application of *units outside the system*, which is permitted by the ISO 31 standard. These include units of the atomic system of units, which are more convenient in solving nuclear physics problems, as well as astronomical units of length and such units as the ton, hour, day, and hectare.

5.3   Derived Units of SI. Dimension of Quantities and Units. Multiple and Submultiple Units

In accordance with Carl Gauss' principle, the derived units must be reproduced by the method of indirect measurements using the materialization of the corresponding coupling equation between units of mutually linked units. In the SI, which is a coherent system of quantities, derived quantities are formed with the aid of coupling equations between quantities in which the numerical coefficients are equal to 1. (*A coherent system of units is a system of units in which all derived units are linked to*

Table 20 Derived SI units that have special designations

Quantity		Unit		
Designation	Dimensionality	Designation	Symbol	Expression in terms of base units of the SI
Planar angle	None	Radian	rad	–
Solid body angle	"	Steradian	sr	–
Frequency	T^{-1}	Hertz	Hz	s^{-1}
Force	LMT^{-2}	Newton	N	$m\ kg\ s^{-2}$
Pressure	$L^{-1}MT^{-2}$	Pascal	Pa	$m^{-1}\ kg\ s^{-2}$
Energy, work, heat, quantity of heat	$L^{2}MT^{-2}$	Joule	J	$m^{2}\ kg\ s^{-2}$
Power	$L^{2}MT^{-3}$	Watt	W	$m^{2}\ kg\ s^{-3}$
Amount of electricity	TI	Coulomb	C	sA
Electrical voltage, potential, EMF	$L^{2}MT^{-3}I^{-1}$	Volt	V	$m^{2}\ kg\ s^{-3}\ A^{-1}$
Electrical capacity	$L^{-2}M^{-1}T^{4}I^{2}$	Farad	F	$m^{-2}\ kg^{-1}\ s^{4}\ A^{2}$
Electrical resistance	$L^{2}MT^{-3}I^{-2}$	Ohm	Ω	$m^{2}\ kg\ s^{-3}\ A^{-2}$
Electrical conductivity	$L^{-2}M^{-1}T^{3}I^{2}$	Siemens	Sm	$m^{-2}\ kg^{-1}\ s^{3}\ A^{2}$
Magnetic induction flux	$L^{2}MT^{-2}I^{-1}$	Weber	Wb	$m^{2}\ kg\ s^{-2}\ A^{-1}$
Magnetic induction	$MT^{-2}I^{-1}$	Tesla	T	$kg\ s^{-2}\ A^{-1}$
Inductance	$L^{2}MT^{-2}I^{-2}$	Henry	H	$m^{2}\ kg\ s^{-2}\ A^{-2}$
Luminous flux	J	Lumen	lm	cd sr
Luminous power	$L^{-2}J$	Lux	lx	m^{-2} cd sr
Radionuclide activity	T^{-1}	Becquerel	Bq	s^{-1}
Absorbed dose of ionizing radiation	$L^{2}T^{-2}$	Gray	Gy	$m^{2}\ s^{-2}$
Equivalent radiation dose	$L^{2}T^{-2}$	Sievert	S	$m^{2}\ s^{-2}$

other units of the system by equations in which the numerical coefficients are taken as equal to 1. For example, to define the derived unit of pressure, the coupling equation between quantities $P = F/S$ is used, where P is pressure caused by force F, equally distributed over the surface; and S is the area of the surface located perpendicular to the force. Let us write this equation in the form of an equation between units: $[p] = [F]/[S]$. Substituting into this equation the units of force, 1 N, and area, $1\ m^{2}$, we derive: $[P] = 1\ N : 1\ m^{2} = 1\ N/m^{2}$. This derived unit is assigned the designation of "pascal" (Pa). Table 20 shows the derived SI units that have special names.

An important concept of systems of units is the dimensionality of the quantity (and, correspondingly, the units of this quantity). The dimensionality of a derived unit indicates to what number of degrees this unit changes with specified changes in the base units. If, when a base unit is changed by a factor of n, the derived unit changes by a factor of n^{p}, it is stated that the derived unit has dimensionality of p relative to the base unit. For example, area has dimensionality of two relative to length, and acceleration has dimensionality of negative two relative to time.

Hence, *a formula for the dimensionality of a quantity is an expression in the form of a power term that is the product of the symbols of base quantities at various powers that reflect the link between this quantity and the base quantities.* Let the symbols of base quantities be: L – length, T – time, M – mass, θ – temperature, I – force of electrical current, J – luminous power, and N – amount of substance. Then in the example discussed above, the dimensionality of force is equal to LM/T^2, and consequently the dimensionality of pressure is equal to $LM/T^2L^2 = M/T^2L$.

Let us note the following properties of dimensionality:

1. We designate the dimensionality of a quantity A as $\overset{\leftrightarrow}{A}$. Then the following properties of dimensionality are evident:

   - if $C = AB$, then $\overset{\leftrightarrow}{C} = \overset{\leftrightarrow}{A} \cdot \overset{\leftrightarrow}{B}$;
   - if $C = A/B$, then $\overset{\leftrightarrow}{C} = \overset{\leftrightarrow}{A}/\overset{\leftrightarrow}{B}$;
   - if $C = A^n$, then $\overset{\leftrightarrow}{C} = [\overset{\leftrightarrow}{A}]^n$.

2. Dimensionality does not depend on the coefficients in equations coupling the quantities. For example, the dimensionality of the area of a circle, $\overset{\leftrightarrow}{S}_{\text{cir}} = \pi/4(\overset{\leftrightarrow}{d})^2 = \overset{\leftrightarrow}{L}^2$, is the same as for the area of a square.

3. Dimensionality reflects the deep links between quantities, which are more general than the links describing physical equations. The same dimensionality can be inherent in quantities with different nature. For example, the work of a force F over a distance l is equal to $A = Fl$. The kinetic energy of a body of mass m moving at velocity v is equal to $B = 0.5mv^2$. The dimensionality of these qualitatively different quantities is identical: $\overset{\leftrightarrow}{A} = \overset{\leftrightarrow}{B} = ML^2/T^2$.

As applied to this property, all quantities are subdivided into dimensional and dimensionless. *A quantity, in the dimensionality formula of which the symbol of even one base quantity is represented with a non-zero exponent, is called a dimensional quantity.* Consequently, force and pressure are dimensional quantities. *A quantity, in the dimensionality formula of which all base quantities have zero exponent, is called a dimensionless quantity.* For example, the molar fraction of a component is a dimensionless quantity, since the numerator and denominator in the formula for determining its dimensionality cancel each other.

Hence the value of a dimensional quantity will always be written with a concrete number, and the value of a dimensionless one with an absolute number (*a number with the designation of its constituent units is called a concrete number*). However, this does not mean that a dimensionless number is not a full-valued physical quantity existing in the physical world but is just a number, a mathematical abstraction. In the definition of a quantity it is stated that a quantity is one of the properties of a physical object. This fully corresponds also to the content of the understanding of a dimensionless quantity. Further, a unit of any dimensionless quantity can be materialized. For example, the molar fraction of a component B is defined by the expression $x_B = n_B/n$, where n_B is the amount of the substance of component B, and n is the amount of substance of the system. Consequently, in

Table 21 Prefixes for multiple and submultiple SI units

		Prefix designation				Prefix designation	
Exponent K	Prefix	International	Russian	Exponent K	Prefix	International	Russian
18	Exa	E	Э	−1	Deci	d	д
15	Peta	P	П	−2	Centi	c	с
12	Tera	T	T	−3	Milli	m	м
9	Giga	G	Г	−6	Micro	μ	мк
6	Mega	M	М	−9	Nano	n	н
3	Kilo	k	к	−12	Pico	p	п
2	Hecto	h	г	−15	Femto	f	ф
1	Deca	da	да	−18	Atto	a	a

order to reproduce the unit of the molar fraction of component B, it is necessary to obtain the pure, unadulterated substance of this component. In fact, in this substance $n = n_B$ and consequently $x_B = 1$.

The SI system also contains multiple and submultiple units. *A unit of a quantity that is a whole number of times greater than the system unit is called a multiple unit.* For example, a unit of length 1 km $= 10^3$ m is a multiple of a meter. *A unit of a quantity that is a whole number of times less than the system unit is called a submultiple unit.* For example, a unit of length 1 nm (nanometer) $= 10^{-9}$ m is a submultiple of a meter. For ease of use, all multiple and submultiple SI units are formed in accordance with the decimal number system; they are equated to the system unit multiplied by a factor equal to 10^K, where the exponent K is a positive number for multiple units and negative for submultiple units. The exponents K and the corresponding prefixes are shown in Table 21.

5.4   Quantities and Units of Physical–Chemical Measurements

Special examination of the quantities and units of physical and chemical measurements in a general course in metrology makes sense, since these measurements are remarkably widely distributed throughout modern society, and they have definite specificity. The most widespread quantities and units of this area of measurements are shown in Table 22.

As seen in Table 22, there is a broad group of various physical quantities that reflect the composition of substances. The convenience of presenting materials in scientific publications, standards documents, and technical reports evoked the need for a generalized designation for the quantities of this group. Thus there appeared the concept of the *component percentage*, by which is understood *the generalized name of a group of quantities that reflect the composition of the substances.* Using the term "concentration of the component" along with it as a generalized concept is impermissible, since a more narrow definition of this concept was compiled long ago.

Table 22 The most widespread quantities and units of physical and chemical measurements

Designation of quantity	Defining equation	Unit
Mass component of component B	$\rho_B = \frac{m_B}{V}$, where m_B is the mass of component B in the system and V is the volume of the system	kg/m^3
Molar concentration of component B	$c_B = \frac{n_B}{V}$, where n_B is the amount of substance of component B in the system	mol/m^3 mol/dm^3
Molecular (atomic) concentration of component B	$C_B = \frac{N_B}{V}$, where N_B is the number of molecules (atoms) of component B in the system	$1/m^3$ $1/dm^3$
Mass ratio of component B	$w_B = \frac{m_B}{m}$, where m is the mass of the system	%
Molar ratio of component B	$x_B = \frac{n_B}{n}$, where n is the amount of substance of the system	%, ‰ ppm, ppb
Volume ratio of component B	$\varphi_B = \frac{V_B}{\sum x_A V_{m,A}^*}$, where V_B is the volume of component B in the system, and $V_{m,A}^*$ are the molar volumes of the individual substances	%, ppm, ppb
Mass ratio of component B	$R_B = \frac{m_B}{m - m_B}$	%
Molarity of component B	$b_B = \frac{n_B}{m_A}$, where m_A is the mass of the solvent	mol/kg
Partial pressure of component B	$P_B = x_B P$, where P is the pressure of the gas mixture	Pa

In L.A. Sena's foundational monograph "Units of Physical Quantities and Their Dimensions" [30] was written: "referring the number of units to the unit of volume, we derive a quantity which is called concentration". This principle was implemented in ISO Standard 31, and in other standards documents: the term "concentration" is generic for physical quantities, characterizing the content of the component per unit of volume. Actually, let us turn to Table 22. It is clear from it that the concepts, except for two specific ones, are divided into two groups: quantities that define the content of a component per unit of volume, and relative quantities. For the first group the concept of "concentration" is the generic concept, and for the second group it is "fraction". The generalized name of all these concepts – content of the component – is not repeated by any of the subordinate concepts. This eliminates the possibility of ambiguous interpretation. If one uses the term "concentration" as a synonym for the concept of "content", then it will turn out that two different concepts, the generalized and the subordinate, will have the same terms. Recognizing the danger in such a confusion of concepts, L.R. Stotsky wrote, in his reference book on physical quantities and units [31]: "Fraction is a dimensionless relative quantity – in no circumstances may one call it concentration".

Let us turn again to the quantities shown in Table 22. It strikes one immediately that for a specific chemical composition and specified test conditions, there is a mutually unambiguous relationship between these quantities, and hence it is possible to execute the conversion of one quantity into the other. For example, one may write a chain of equalities:

$$\frac{m_B}{V} = \frac{m_B}{m}\frac{m}{V} = \frac{n_B}{V}\frac{m_B}{n_B} = \frac{V_B}{V}\frac{m_B}{V_B}$$

Since $m/V = g$ and $m_B/V_B = g_B$ are the density of the system and of component B, but $m_B/n_B = M_B$ is the molar weight of component B, there follow from these equalities the relationships between mass concentration ρ_B, mass fraction ω_B, molar concentration c_B and volumetric fraction ϕ_B of component B:

$$\rho_B = \omega_B g = c_B M_B = \phi_B g_B.$$

The interconversion of these quantities is widely used in physical and chemical measurements. This practice can lead to the opinion that in chemical analysis there is only one quantity that characterizes the composition of an object – namely the composition itself – and that the quantities presented in Table 22 are just different ways of expressing this quantity [32]. However, these are different concepts. *Composition is the aggregate of some elements entering some chemical compound or substance as constituent parts, and a property is a quality or attribute constituting the distinguishing features of something* [33]. The difference between these two concepts is obvious. The same composition of an object has different distinguishing attributes, and is characterized by different properties and, consequently, quantities. For example, the mass fraction characterizes the ratio of the mass of the component and of the object, and the volumetric fraction characterizes the ratio of their volumes. It is clear that this is not the same ting, especially considering that the former does not change when the temperature of the object changes, while the latter does change.

From the metrological point of view, non-homogeneous quantities can be considered identical if the unit of any of these quantities can be derived computationally from the unit of another quantity without increasing the uncertainty. Writing a quantity in any of the multiple or submultiple units satisfies this requirement. If, in converting one unit into another, the values of other quantities such as fundamental physical constants are used, then the uncertainties of these values are added to the uncertainty of the first unit. In doing so, the uncertainty of the second unit becomes larger than the uncertainty of the initial unit. Consequently, these units are not identical, and as a result of the conversion we have transitioned to a different quantity.

Let us introduce the example of wavelength and frequency. These quantities are not identical, since their interconversion is based on a fundamental physical constant – the Planck constant, known with some uncertainty. Hence they are dissimilar quantities. Precisely the same way, the various properties that reflect the chemical composition of substances are dissimilar quantities.

Chapter 6
Measurement Assurance

6.1 Uniformity and Traceability of Measurement

Measurements have significant meaning in any area of human activity. But in order for measurements to be effectively used in practical activity, one must be convinced that their results remain unchanged (within the desired accuracy) when using another instrument and measurement method, and when changing the operator, time, and place of completion of measurements, and other components of measurement processes. This property of measurements is called uniformity of measurements in Russia and other Commonwealth of Independent States (CIS) countries.

To understand the significance of uniformity of measurements, let us look at an example of one type of measurement. Let us suppose that the state's unified time service ceased to function, and all enterprises and citizens of the country began to relay upon their own watches, not calibrating them by accurate time signals. After a short time there would be many problems associated with beginning the workday, with the traffic schedule of various kinds of transport, and with the operating time of stores and enterprises providing the population's utility services. For example, a passenger goes into a train station, and they tell him "your train has already left." In time, the number of similar cases would abruptly rise, and in the end would lead to the collapse of the entire transport system of the country. Untimely delivery of equipment, raw materials and component parts, feed and other products would lead to disorganization of the work of enterprises and to the well-founded refusal of customers to pay for such deliveries, leading in the end to a crisis in the financial system of the country. Without knowledge of a unified time, the operation of communications and navigational systems, health and education agencies, etc., would be impossible. It is clear that the functioning of a state service for unified time is a necessary condition for the existence of any state.

The very same thing can be said about other forms of measurement. Uniformity is the main and determining mark of measurements, which in principle differentiates them from other kinds of experimental evaluation. If uniformity is

A.E. Fridman, *The Quality of Measurements: A Metrological Reference*,
DOI 10.1007/978-1-4614-1478-0_6, © Springer Science+Business Media, LLC 2012

ensured, then this means that the process of experimental evaluation being studied
can pertain to measurements. If it is not ensured, then this suggests the unsuitability
of this process as measurement, which it consequently is not. In this case, it is some
other process of recognition – observation, organoleptic evaluation, indication, and
so forth – but not measurement. Hence by virtue of its importance, measurement
assurance is the chief goal and foundational content of the theoretical and practical
work of metrologists.

What is understood by the term "uniformity of measurement"? In accordance
with the classical definition provided in [9], *uniformity of measurement is the state
of measurement in which all results are expressed in authorized units, and all errors
are known with a specified probability*. This definition, formulated in the 1960s by
Prof. K.P. Shirokov, has become significantly obsolete. First, the condition that "all
results are expressed in authorized units" was included in the definition according to
considerations of the time, since at that time the inculcation of the SI International
System of Units was the most pressing task of Russian metrologists. By now, issues
regarding the introduction of the SI have long been fully resolved in our country and
in many others. In that same regard, the experience of the USA and other English-
speaking countries has demonstrated that mass usage in home, trade, and industry
of traditional (for these countries) units that are forbidden by the SI (pound, foot,
degrees Fahrenheit, etc.) does not have serious negative consequences.

The second condition that "all errors are known with a specified probability,"
corresponding in sense to the interpretation of uniformity of measurement but not
being specific, was formulated otherwise in the succeeding terminological
standards document [34]. However, even the new terminology that "measurement
error shall not exceed established bounds" is a declaration, expressing the goal of
metrology but in practice unachievable (as the true value of a quantity is unknown).
Since, due to the random nature of measurement error in any measuring technology,
the probability of a measurement defect is greater than zero, there will always be
found a measurement, even if just one, whose error exceeds the established bounds.
Consequently, if we interpret this definition literally, uniformity of measurement
never was and never will be.

So what sense ought we to imbue the concept of "uniformity of measurement"? In
answering this question let us consider that in order for measurement results not to
depend on the selection of measuring instruments [MIs], the MI equipment must
satisfy the following condition: **all measuring systems of the same quantity must
preserve the same unit size.** A natural way of fulfilling this condition consists of
reproducing a unit with one reference standard and subsequent transfer of this unit size
to all other measuring systems for that quantity. From this comes another
term accepted in international practice, reflecting the concept under study –
*measurement traceability. This is understood to mean the fact that a measurement
result is obtained by comparing the size of a measurand with the size of a unit,
reproduced using a reference standard of that quantity* (literally: a measurement
result is traced to the reference standard). This signifies that they are done as if by
reference to one MI – the reference standard. It is evident that uniformity of measurement is a consequence of their traceability, and this provides the basis for considering

the terms as synonyms. In this regard, the new definition is better than the classical one and clarifies the essence of the concept "uniformity of measurement." Hereafter we shall use the more convenient term – uniformity of measurement.

6.2   Measurement Assurance

The task of ensuring uniformity of measurement has stood since time immemorial, since the necessity for doing so arose simultaneously with the need for measurements. With that, to the extent that science and industry developed, the level of requirements for uniformity of measurement also grew continuously. This is reflected comprehensively:

- it encompasses an ever greater number of areas and kinds of measurement, and the ranges of measurement continuously increase;
- all spheres of application of measuring technology are encompassed (trade, various sectors of industry and agriculture, warfare, science, health care, transport, communications, environmental protection, and so forth);
- the level of measurement accuracy and the requirements for measurement accuracy are constantly increasing;
- the ground covered by uniformity of measurement is broadening – in the Middle Ages a separate market or enterprise, later a settled point, then a region, the state, and finally the entire world;
- the significance of uniformity of measurement is increasing and, consequently, the material, financial and labor resources selected to ensure it.

Towards the end of the twentieth century, the system built up over the ages (we call it the classical system) to ensure uniformity of measurement was definitively formulated. At its base lies the definition of traceability as presented in Sect. 1. To ensure uniformity in any form of measurement, it is essential to:

- develop in a collegial manner, and authorize, a theoretical definition of this quantity,
- reproduce this unit by some defined standard as a reference, with the highest accuracy possible,
- regularly, and with a periodicity necessitated by the instability of the MI, transfer the unit size, as preserved by the reference standard, to all MIs for this quantity, likewise with the highest accuracy possible.

To resolve the first issue, the most authoritative physical and metrological laboratories of the world conduct scientific research directed toward recognizing new physical governing laws and refining the fundamental physical constants. Based on the results of these studies, the highest collegial body of the world metrological community – the General Conference on Weights and Measures (CGPM) – periodically authorizes new theoretical definitions of the units of physical quantities.

To resolve the remaining issues, hierarchical systems are created:

- standard samples, in which all except the reference standards are subordinated to other, more accurate standard samples,
- metrological documents that regulate the reproduction of units and the transfer of their dimensions,
- metrological services performing these operations.

In Russia, the first of these named systems is called the technical base of the measurement assurance system, the second is called the standards base, and the third is called the technical base.

In industrially developed states, the hierarchical systems of standards for the majority of the types of measurement form centralized systems for reproducing units and transferring their dimensions. These systems include reference standards for the country, called national standards, and hierarchical chains of subordinate standards, which execute the transfer of unit size from the national standard to all MIs for that quantity that are used in the country. By way of exception, in a number of types of measurements, there are found decentralized systems of measurement assurance in which there are no national standards. In this case, the unit is reproduced in one of two ways:

- using several reference standards located in various laboratories, where the unit size is transferred to the reference standards in foreign metrological centers;
- by the method of indirect measurement, directly in measurement laboratories.

The centralized system is capable of ensuring the highest level of uniformity of measurement. Hence as a rule the decentralized systems are used at the initial stages of developing a form of measurement. The completion of any type of measurement proceeds by developing methods and instruments for comparative measurements and creating standard samples, based on which hierarchical chains of standards arise. The natural endpoint of this process is the creation of a national standard and a centralized system of reproducing a unit and transferring its dimension.

6.3 Standards of Units

It follows from Sect. 2 that in Russia and many other countries of the world, uniformity of measurement is ensured by the functioning of centralized systems of reproducing units of quantities and transferring their dimensions, which are hierarchical chains of standards of various accuracy. *A standard is a measuring system (or set of measuring systems) designed to reproduce and (or) preserve a unit and transfer its size to other measuring systems.* Standards are subdivided into primary, secondary and operational according to their position in this hierarchical chain.

A standard that reproduces a unit and transfers its size to other standards is called a primary standard. A primary standard executes the task of reproducing a unit of a quantity for use in all measurements of that quantity. It is evident that the

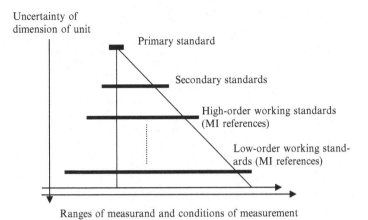

Fig. 13 Framework for reproducing a unit and transferring its size to standards

levels of accuracy of the most authoritative metrological and operational measurements are determined by the accuracies of the primary standards. Hence in the creation of primary standards, the attempt is always made to ensure the highest accuracy that can be achieved at this stage of development of science and technology. After reproduction of a unit, its size is transferred to each standard following the hierarchical chain.

Figure 13 illustrates this process. It is clear from the figure that transfer of the unit size occurs in two ways: not just from the most accurate standards to the least accurate but also by broadening the ranges of the quantity and conditions of measurement. Here, since the result of each measurement is burdened with some uncertainty, the uncertainty of this unit continuously increases in the system of transferring the unit size. The primary standard transfers the unit size to the secondary standards, which function in a wider range of measurements but are less accurate. The secondary standards transfer the unit size to the operating standards [measuring instrument standard references (hereafter SI references)], and these to less accurate working standards.

The number of steps in the transfer is determined by the requirements for accuracy of the working measuring instruments, and hence cannot be very large. In many types of measurement, an increase in the ranges of the quantity and the conditions of measurement (frequency, temperature, etc.) has made it impossible to ensure the transfer of the unit size with the required accuracy from the operative primary standard to all measuring instruments of this type. In these cases, several primary reference standards of one unit are created, differing by the ranges of measurement or conditions of measurement. For example, in Russia, the unit of pressure is reproduced by seven different primary reference standards, six of which are kept at the D.I. Mendeleev VNIIM [All-Russian Scientific Research Institute for Metrology] and one at the VNIIFTRI [All-Russian Scientific Research Institute for Physico-Technical and Radio Measurements]. Another example: the unit of specified heat capacity is

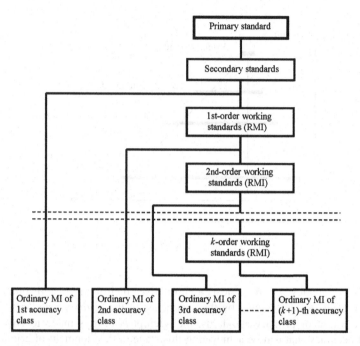

Fig. 14   Transfer of dimensions of unit from reference samples to ordinary MI

reproduced from four primary standard references kept at the D.I. Mendeleev VNIIM, UNIIM [Urals Scientific Research Institute for Metrology], NPO [Scientific Design Organization] Dal'standart, and VNIIFTRI. This group of primary standards is, in fact, a collective standard designed to reproduce the unit throughout the whole range of measurements. In this, one of these standards in Russia receives the designation of "primary standard." The other standards of this set are called special primary standards (short form: "special standards"). In the examples given, the designation of primary standards is given to state reference standards stored at the D.I. Mendeleev VNIIM, for units of pressure in the range from 0.05 to 10 MPa and specific heat capacity between 273.15 and 700 K. To ensure uniformity of measurements, the unit sizes reproduced by these standards are made to conform at the boundary regions of the values.

All other standards do not take part in the reproduction of units. They preserve the dimensions of units obtained from more accurate standards, and (or) transfer them to less accurate standards and working MIs. Figure 14 illustrates the transfer of dimensions of units from standards to working MIs. *Secondary standards receive the unit sizes from primary standards and transfer them to working standards. Working standards are designed for verification and calibration of working MIs.* When necessary they are subdivided into orders: first, second, third, etc. In this case, the first-order working standards likewise transfer the unit sizes to second-order working standards, and second-order working standards likewise transfer the unit sizes to third-order working standards, and so forth.

The designations introduced above correspond to the international classification of standards that is recognized throughout the world, including Russia. At the same time, the classification adopted in Russia differs somewhat from the international. The working standards in the international classification were called MI reference standards in our country [Russia] until the last decade. These terms are in fact synonyms: in its own time the French word "etalon" [reference standard] was transferred into Russian. But in Russia, there were always substantive differences in the status of primary and secondary standards (the combination of which is called the standards base of the country), and the OSI and the differences caused by it in the financing and the system of metrological service. After the transition to the international classification, the status of the OSI (including serially issued OSIs for lower orders) was as though compared with the status of primary and secondary standards, with all the resulting consequences of a negative nature for the latter. To reestablish the former position and at the same time not contradict the international classification of standards, a decision has currently been adopted to use both terms ("working standard" and "OSI") as being equally valid.

The standards include:

- means of reproducing a unit (primary measuring transducers, measurement installations);
- means of preserving unit dimensions (measures);
- means of transferring unit dimensions (comparators, comparison standards);
- means of preserving and transferring unit dimensions (measuring instruments);
- other MIs and technical facilities (means of monitoring measurement conditions, computational facilities, power grid, measuring equipment accessories, and other).

In design, the standards and OSI can be formulated as measuring installations, called verification installations in this case. The basic metrological requirements for standards and OSI must ensure high accuracy of measurement results when reproducing a unit and preserving and (or) transferring its dimension. The most important requirement is high stability enabling one to ensure the fixed nature of the unit dimension. It is exactly this that explains the transition that took place at the beginning of the twenty-first century toward reproduction of the base and many derived units by implementation of high-stability quantum physical effects. High sensitivity and small random error of the standard are also necessary, as well as its low sensitivity to change in the measurement conditions.

Automation of measuring and computational operations is the way to improve measurement accuracy at the standard and to reduce personal operator error. At the same time, the extent of the range of measurement conditions is not a significant characteristic since, for the purpose of improving the accuracy of standards measurements, the standards are as a rule used in fixed (most commonly normalized) measurement conditions.

Individual standards, group standards, and collective standards are differentiated based on the number of similar MIs in the standard.

An individual standard consists of one MI. Another form of standards, the group standard, is a set of individual standards with differing nominal values, which permits broadening the range of reproduced, preserved, and (or) transferred dimensions of units. For example, the state primary standard of the unit of temperature (GPE) is a group standard. It consists of two standards, one of which, kept at VNIIFTRI, reproduces the unit of temperature for the range 0.3–273.16 K, and another, kept at the D.I. Mendeleev VNIIM, does so for the range 0–3,000°C. Each of these standards is likewise a group standard, since it consists of a set of facilities to reproduce the reference points of the ITS-90 International Temperature Scale and a set of thermometers and pyrometers for measuring the temperatures at these points [35].

A collective standard is a group of standard reference MIs (most commonly the measures of one nominal value), united with the goal of improving the stability and accuracy of the standard. The effect of creating a collective standard is explained as follows. One may present as the following sum the value x_i of the ith standards measure at moment t, calculating from the time of the last calibration of this standard:

$$x_i = \dot{x}_i + \xi_{Ai} + \xi_B + v_i(t), \qquad (6.1)$$

where $\dot{x}_i = \dot{x} + \Delta x_i$ is the actual value of the ith measure ($\dot{x}$ is the nominal measure of the collective standard,

Δx_i is the error in preparation of the ith measure),

ξ_{Ai}, ξ_B are the measurement uncertainties in calibration (transfer of the unit size) of the ith measure of types A and B, respectively ($\ddot{x}_i = \dot{x}_i + \xi_{Ai} + \xi_B$ is the value assigned to the measure upon calibration), and

$v_i(t)$ is the change, over time t after calibration, of the actual value of the $\dot{x}_i$ of the ith measure.

If this measure is applied as an individual standard, its uncertainty will be defined as the sum of three terms ($\xi_{Ai} + \xi_B + v_i(t)$). The standard uncertainty of such a standard will be equal to

$$u_i = \sqrt{u_A^2 + u_B^2 + u_v^2(t)}, \qquad (6.2)$$

where u_A, u_B are the standard uncertainties of the results of measuring the value of a measure at calibration, and evaluated as type A and B; and

$u_v(t)$ is the standard uncertainty of the change in value of the measure over time t.

Now, let us join into one collective standard n such measures having the same nominal value $\dot{x}$. The value of this standard is the mean of the values of all n measures:

$$\bar{x} = \frac{1}{n} \sum_{i=1}^{n} x_i = \dot{x} + \bar{\Delta}x + \bar{\xi}_A + \xi_B + \bar{v}(t), \qquad (6.3)$$

where $\Delta x = 1/n \sum_{i=1}^{n} \Delta x_i$ is the mean manufacturing error for the measures of the collective standard,

$\bar{\xi}_A = 1/n \sum_{i=1}^{n} \xi_{Ai}$ is the mean uncertainty of the results of measuring the values of the measures at calibration of the collective standard, evaluated as type A, and

$\bar{v}(t) = 1/n \sum_{i=1}^{n} v_i(t)$ is the change in value of the collective standard over time t, equal to the mean change over that time of the values of the measures included in this standard.

With that, the actual value of the collective standard is equal to $\bar{x} = 1/n \sum_{i=1}^{n} \dot{x}_i = \dot{x} + \Delta x$, the value assigned to the collective standard at calibration is $\bar{\bar{x}} = 1/n \sum_{i=1}^{n} \ddot{x}_i = \bar{x} + \bar{\xi}_A + \xi_B$, and the value of the collective standard at time t after calibration is $\bar{x} = 1/n \sum_{i=1}^{n} x_i = \bar{\bar{x}} + \bar{v}(t)$.

Usually measures of one type are collected into the collective standard. With that being the case, changes over time t of the values of the measures incorporated into the collective standard are individually distributed mutually independent random quantities. Due to this, the standard uncertainty of the instability of the collective standard, as the standard deviation of the mean, is equal to

$$u[\bar{v}(t)] = \frac{u_v(t)}{\sqrt{n}}.$$

Hence, the stability of the collective standard, including n measures on one type, exceed by a factor of $\sqrt{n}$ the stability of any one measure of this standard. The accuracy of the collective standard will also be higher: the standard uncertainty of its value at time t is equal to

$$u_{\Gamma.э} = \sqrt{\frac{u_A^2}{n} + u_B^2 + \frac{u_v^2(t)}{n}}. \tag{6.4}$$

Comparison with formula (6.2) illustrates the effect of using collective standards in systems for transferring dimensions of units.

A centralized system of measurement assurance can be organized in the metrological services of diverse level: at a separate enterprise, in a department, and in the country as a whole. Here, the one most accurate standard will be at the head of this hierarchical chain of transfer of the unit size to all MIs of this type of measurement that are used in this metrological service. This standard is called the reference standard (of the enterprise, department, or country). Hence traceability is ensured for all measurements performed at the enterprise or department back to its reference standards, and through them to the country's reference standards.

The reference standards of a country in international metrological practice are called national standards, and in our country [Russia] they are called state standards. National standards, being more accurate than the MIs of their own countries, greatly determine their scientific and technical capabilities. They are usually preserved and used in national metrological institutes (NMIs). There are three different methods for creating them [36].

In accordance with the first method, an NMI creates a primary standard which performs the reproduction of a unit in precise correspondence with its definition.

This is the most fundamental approach, which ensures a strong link between the definition of a unit and its physical embodiment in the primary standard. However, it is the most difficult and expensive approach.

The second method likewise calls for the creation of a primary standard, although not by means of the physical implementation of a theoretical definition, but based on quantum physical effects the use of which makes it possible to implement a unit size that corresponds with its definition. Such an approach facilitates creation of standards with a high degree of reproducibility. Examples are the use of frequency-stabilized lasers to create the meter standard, the Josephson effect for the volt, and the Hall effect for the ohm.

The third method involves the use, for the national standard, of a secondary standard to which the unit size is transferred by periodic comparison with the national standard of some other NMI.

The largest and most scientifically and technologically developed countries, including Russia, having the need for the highest possible level of accuracy in all types of measurement, use the first and second methods. Hence, the national standards of these states, as a rule, are primary standards. Small countries, not having the full spectrum of modern sectors of industry and hence not sensing the need for high accuracy in a number of types of measurement, are often required to use the third method to create national standards, as being less expensive. In considering this, the BIPM has established that each NMI shall have the right, in creating its national standard, to independently select the method of its creation.

6.4 Verification and Calibration of Measuring Equipment

The procedure for transferring a unit size from a standard to a less accurate MI can be done in two different forms – calibration and verification.

Calibration is the simplest form and includes just the transfer of unit size. *Calibration is the name given to the set of operations that establish a relationship between the value of the quantity obtained using the given MI, and the corresponding value of the quantity determined using the standard.* It follows from this definition that in calibration there is a transfer of the quantity's unit size, reproduced or preserved by the standard, to a less accurate MI by means of determining the ratio between the values of the quantity preserved by the standard and the corresponding indicators of the MI being calibrated. In the preceding, when using the MI for its purpose, this ratio is used to transform MI indicators into measurement results. In the process of calibration, the measures are evaluated and assigned new values, and the measuring instruments are assigned new calibration characteristics (*a calibration characteristic is a dependence between the output signal y of the measuring instrument and the measurand x*). The value of the measure, determined in calibration, is indicated by the assigned value or by an additive correction equal to the deviation of this value from the value assigned to the measure at the original or succeeding calibration. The calibration characteristic of the measuring instrument or measuring transducer, determined at calibration, is shown in the form of a new function $x = f(y)$ or $y = f(x)$ (by a function, table, or graph). If the calibration characteristic is linear ($y = A + Kx$), the new characteristic can be indicated by the corrections (additive and multiplicative) to the assigned calibration characteristic. *A correction not dependent on the measurand is called an additive correction, and a correction directly proportional to the measurand is called a multiplicative correction.* It is evident that an additive correction indicates the new value of A, and a multiplicative correction the new value of K. If the design of a measuring instrument being calibrated permits resetting to zero before use, then one may consider that $A = 0$. In this case, the calibration characteristic is provided as a calibration coefficient K, which is constant for all points of the range (or part of the range) of a measuring instrument. It is evident that in the calibration process the systematic errors of the MI that have accumulated to this time will generally be eliminated, no matter how significant, and the MI will be reestablished with its initial metrological characteristics. Hence an MI cannot be rejected at calibration in principle. The purpose of calibration is not to evaluate the compliance of the MI with the established technical requirements. That question is resolved with verification.

Verification is the establishment, by an agency of the state metrological service, of the suitability of the MI for use, on the basis of test analysis of its errors. It is clear from this definition that verification consists of two different procedures: technical and directional. The technical procedure consists of experimental evaluation of the MI's errors. Since in measuring the values x of the quantity the error of the MI is equal to $\Delta(x) = \hat{x} - x$, where $\hat{x}$ is the result of measuring this quantity by the MI being verified, and x is the value of the measurand determined with the aid of the standard, it is understood that the technical procedure is in principle analogous to the procedure of calibrating the MI. The directional procedure consists of rendering a decision on the compliance of the MI being verified with the set of requirements, expressed in the form of bounds of allowable error values $\pm \Delta_\partial(x_i)$, and the formulation of a document as to the allowance or disallowance of the MI for further use. When the condition $-\Delta_\partial(x_i) \leq \Delta(x_i) \leq \Delta_\partial(x_i)$ is satisfied at all control points x_i of the range of measurements, the verifying agency shall issue to the owner of the MI a certificate permitting the further use of the MI, but if this condition is violated at even one point x_i, use is forbidden. It is interesting that in some countries this separation of functions has led to its logical conclusion – the specified procedures are executed by different services: the experimental evaluation of errors by calibration laboratories, and the rendering of a decision on compliance, called a verification, by state verification agencies, based on the certificate of the calibration laboratory.

This fact underscores the distinction between the legal status of calibration and verification. Verification is one of the forms of state metrological control, which is extended to measurements performed for: defense of human life and health, environmental protection, and execution of trade operations and mutual accounts between buyer and seller. The list of MIs subject to verification is confirmed by the Government of the Russian Federation. Hence there are a series of strict requirements:

- inadmissability of the use, for measurements subject to state metrological control and supervision, of MIs not verified or that are overdue for the next verification;

- verifications to be conducted exclusively by agencies of state metrological control and supervision or by organizations fully authorized by these agencies;
- strict observance of the requirements of the verification methodology confirmed at the state level and of the established intervals between verifications.

All other MIs are subject to calibration. In contrast with verification, calibration is not obligatory. It is performed at the wish of an organization that used an MI, using its own resources or specialized calibration laboratories on a commercial basis, and employing a calibration method agreed upon with the organization that is using the MI. The fact that there are no essential requirements for calibration controlled by the state is logical, since any harm from invalid measurements of such MIs would be primarily borne by the organizations using them. At the same time, in the conditions of the globalization of the modern market for products and services, an essential requirement for production is documentary confirmation of the traceability of measurements performed in the production process. This confir- mation will provide calibration certificates issued by calibration laboratories that are accredited in the European and Russian system of calibration. Hence enterprises using MIs will, as a rule, prefer to ensure a high quality of calibration, performing it in accredited calibration laboratories and with confirmed methods, observe the set interval between calibrations, and preserve the properly completed certificates of the calibrations.

Calibration ensures higher accuracy of measurements than usual verification, since reestablishment of the initial accuracy of the MI is performed. With usual verification, this is not done – an MI recognized as suitable will be sent out for further use with all the errors that it had when delivered to the verification labora- tory. This simplified algorithm of verification is approved for trade weights and balances, control panel electrical measuring devices, and many other MIs of lower accuracy that were calibrated at the initial verification at the manufacturer, since this makes verification and use of the MI significantly easier. If MIs subject to verification are used for high-accuracy measurements, then their verification, while preserving all inherent legal features (performance by agencies of state metrologi- cal control and supervision, necessity of observing intra-verification interval, etc.) is conducted with the reestablishment of the original accuracy, i.e., using the calibration method.

6.5   Increase of Uncertainty of the Unit Size During its Transfer to a Measuring Instrument Under Calibration or Verification

As was shown in Sect 2, in the hierarchical chain of standards for any form of transfer of unit size, both in calibration and in verification, the uncertainty of unit size increases. But this happens in different ways.

A calibration result will be burdened not just with the uncertainty of the unit size preserved by the standard, but also by many other components. The standard uncertainty of the value of a unit received by an MI in calibration with uncorrelated components will be expressed by the formula

$$u = \sqrt{u_s^2 + u_A^2 + \sum_{i=1}^{n} u_{Bi}^2},$$ (6.5)

where u_s is the standard uncertainty of the value of the unit, reproduced or preserved by the higher-ranking standard used in calibration,

u_A is the standard uncertainty of measurements, and is evaluated as type A, the source of which are random errors of the standard, the MI being verified, and the method of measurement,

u_{Bi} is the standard uncertainty of measurements, and is evaluated as type B, due to the ith source of error in transferring unit size; and

n is the number of sources of transfer error.

For example, in calibrating a measuring instrument by the direct measurements method, the calibration characteristics are determined based on the results of multiple measurements, by the instrument being calibrated, of quantities reproduced by standard measures. The sources of uncertainty of the calibration characteristics are, first, uncertainties of standards measures due to instability of these measures and to the uncertainties of the results of their calibration. But in addition, one should add to these sources of uncertainty: random error of the measuring instrument being calibrated; the nature of influencing values that determine the conditions of measurement in calibration (temperature, pressure, humidity, etc.), from their nominal values; error in formation of calibration characteristics by the least squares method; rounding of measurement results, and so forth. Hence always $u > u_s$, i.e., the accuracy of any MI after calibration is always lower than the accuracy of the standard used in calibrating this MI.

The mechanism by which uncertainty of the unit size increases during certification is totally different. Since verification is in essence a procedure for checking the MI for compliance with the acceptance tolerance Δ, a troublesome occurrence such as statistical errors is significant.[1] Let us return to the theory of statistical errors in verifying an MI. We shall designate as x the error of the MI being verified, and by y the measurement error in verification. As a consequence of measurement error, the result of measurement error in verifying the MI will be equal to $x + y$. The MI is suitable when the condition $-\Delta \leq x \leq \Delta$ is satisfied, where Δ is the bound of allowable error of the MI being verified. But in verification, the MI will be recognized as suitable when the condition $-\Delta \leq x + y \leq \Delta$ is satisfied, which

---

[1] These errors are called statistical to differentiate them from errors caused by violation of the measurement method or operator failure. By contrast with those, statistical errors cannot be assigned to the fault of the operator since they are an objective consequence of measurement uncertainty in verification.

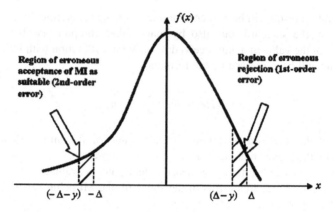

Fig. 15 Regions of values of MI error at which statistical verification errors occur, for $y > 0$

one may rewrite as $-\Delta - y \leq x \leq \Delta - y$. Verification errors are caused by the differences in these conditions. For $y > 0$, a suitable MI will be erroneously rejected (first-order error) if $\Delta - y < x \leq \Delta$, and a defective MI will be erroneously accepted as suitable (second-order error) if $-\Delta - y < x \leq -\Delta$. For $y < 0$, conversely, the first-order error will be when $-\Delta \leq x < -\Delta - y$, and the second-order error for $\Delta \leq x < \Delta - y$.

Figure 15 presents an illustration of this effect. It shows the probability density $f(x)$ of MI error in the set of MIs of one type that are in use, and the intervals of values of the error x at which statistical errors of verification occur. Since the probability is always greater than zero that in this set there are some MIs for which the probability of error lies in these intervals, the occurrence of statistical errors of verification is unavoidable. It is not possible to completely eradicate them, and one can only strive to decrease their share of the total mass of verification results.

Hence, *a first-order statistical error consists of the fact that an actually suitable MI is erroneously rejected due to measurement error in verification. A second-order statistical error consists of the fact that a defective MI is erroneously accepted as suitable due to measurement error in verification.*

Let us estimate the first-order P_1 and second-order P_2 probabilities of statistical error of verification of when verifying MIs, the random error of which are insignificant in comparison with systematic error. Let $f(x)$ be the probability density of error x of the MI being verified in the set of MIs of this type, and $g(y/x)$ the conditional probability density of measurement error when verifying y.

A first-order error occurs when the conditions $-\Delta \leq x \leq \Delta$, and $y < -\Delta - x$ or $y > \Delta - x$, are simultaneously satisfied. In accordance with the formula for the complete probability, we express P_1 with the following integral

$$P_1 = \int\limits_{-\Delta}^{\Delta} \int\limits_{-\infty}^{-\Delta-x} f(x)g(y/x)dydx + \int\limits_{-\Delta}^{\Delta} \int\limits_{\Delta-x}^{\infty} f(x)g(y/x)dydx. \qquad (6.6)$$

A second-order error occurs when the conditions: $x< -\Delta$ or $x>\Delta$, and $-\Delta - x \leq y \leq \Delta - x$, are simultaneously satisfied. Hence the probability of a second-order error is equal to

$$P_2 = \int\limits_{-\infty}^{-\Delta}\int\limits_{-\Delta-x}^{\Delta-x} f(x)g\left(\frac{y}{x}\right)dy\,dx + \int\limits_{\Delta}^{\infty}\int\limits_{-\Delta-x}^{\Delta-x} f(x)g\left(\frac{y}{x}\right)dy\,dx. \qquad (6.7)$$

Measurement error during verification does not depend on the systematic error of the MI being verified. Hence $g(y/x) = g(y)$ does not depend on x, and formulas (6.6) and (6.7) are simplified:

$$P_1 = \int\limits_{-\Delta}^{\Delta} f(x)[G(-\Delta-x)+1-G(\Delta-x)]\,dx,$$

$$P_2 = \int\limits_{-\infty}^{-\Delta} f(x)[G(\Delta-x)-G(-\Delta-x)]\,dx + \int\limits_{\Delta}^{\infty} f(x)[G(\Delta-x)-G(-\Delta-x)]\,dx, \qquad (6.8)$$

where $G(z) = \int\limits_{-\infty}^{z} g(y)\,dy$ is the distribution function of measurement error when verifying.

Usually in estimating statistical errors of verification, it is accepted that the distribution of systematic errors of an MI in a set of MIs of one type is subject to the normal law with zero mean:

$$f(x) = \frac{1}{\sqrt{2\pi}\sigma}e^{-x^2/2\sigma^2},$$

where σ is the standard deviation of this distribution. Likewise it is accepted that the distribution of measurement error in verification is subject to the normal law. Further, let us examine two different ways of posing the problem.

1. Estimate of statistical errors of verification when using a material standard. In this case the measurement error is $y = m + \xi$, where m is the systematic measurement error and ξ is the random measurement error. For standard uncertainty u, estimated as type A, the probability density y is equal to

$$g(y) = \frac{1}{\sqrt{2\pi}u}e^{-(y-m)^2/2u^2}.$$

Further,

$$G(z) = \frac{1}{\sqrt{2\pi}u}\int\limits_{-\infty}^{z} e^{-\frac{(y-m)^2}{2u^2}}\,dy = \frac{1}{\sqrt{2\pi}}\int\limits_{-\infty}^{\frac{z-m}{u}} e^{-t^2/2}\,dt = \Phi\frac{(z-m)}{u},$$

where $\Phi(z) = \frac{1}{\sqrt{2\pi}} \int_{-\infty}^{z} e^{-t^2/2}\, dt$ is the distribution function of a standardized normal quantity (tabulated function); i.e. a quantity whose probability density is equal to $\varphi(t) = 1/\sqrt{2\pi}\, e^{-t^2/2}$. Substituting these expressions into (6.8), we derive:

$$
\begin{aligned}
P_1 &= \int_{-\Delta}^{\Delta} f(x)\left[\Phi\left(-\frac{\Delta + x + m}{u}\right) + 1 - \Phi\left(\frac{\Delta - x - m}{u}\right)\right] dx \\
&= \int_{-\Delta/\sigma}^{\Delta/\sigma} \varphi(t)\left[\Phi\left(-\frac{\Delta + t\sigma + m}{u}\right) + 1 - \Phi\left(\frac{\Delta - t\sigma - m}{u}\right)\right] dt \\
&= \int_{-k}^{k} \varphi(t)\left[\Phi\left(-\frac{k + t + \mu}{\alpha}\right) + 1 - \Phi\left(\frac{k - t - \mu}{\alpha}\right)\right] dt \\
&= \int_{-k(1+\beta)}^{k(1+\beta)} \varphi(s - \beta k)\left[\Phi\left(-\frac{k + s}{\alpha}\right) + 1 - \Phi\left(\frac{k - s}{\alpha}\right)\right] ds,
\end{aligned} \tag{6.9}
$$

$$
\begin{aligned}
P_2 &= \int_{-\infty}^{-\Delta} f(x)\left[\Phi\left(\frac{\Delta - x - m}{u}\right) - \Phi\left(\frac{\Delta + x + m}{u}\right)\right] dx \\
&\quad + \int_{\Delta}^{\infty} f(x)\left[\Phi\left(\frac{\Delta - x - m}{u}\right) - \Phi\left(-\frac{\Delta + x + m}{u}\right)\right] dx \\
&= \int_{-\infty}^{-\Delta/\sigma} \varphi(t)\left[\Phi\left(\frac{\Delta - t\sigma - m}{u}\right) - \Phi\left(-\frac{\Delta + t\sigma + m}{u}\right)\right] dt \\
&\quad + \int_{\Delta/\sigma}^{\infty} \varphi(t)\left[\Phi\left(\frac{\Delta - t\sigma - m}{u}\right) - \Phi\left(-\frac{\Delta + t\sigma + m}{u}\right)\right] dt \\
&= \int_{-\infty}^{-k} \varphi(t)\left[\Phi\left(\frac{k - t - \mu}{\alpha}\right) - \Phi\left(-\frac{k + t + \mu}{\alpha}\right)\right] dt \\
&\quad + \int_{k}^{\infty} \varphi(t)\left[\Phi\left(\frac{k - t - \mu}{\alpha}\right) - \Phi\left(-\frac{k + t + \mu}{\alpha}\right)\right] dt \\
&\quad \int_{-\infty}^{-k(1-\beta)} \varphi(s - \beta k)\left[\Phi\left(\frac{k - s}{\alpha}\right) - \Phi\left(-\frac{k + s}{\alpha}\right)\right] ds \\
&\quad + \int_{k(1+\beta)}^{\infty} \varphi(s - \beta k)\left[\Phi\left(\frac{k - s}{\alpha}\right) - \Phi\left(-\frac{k + s}{\alpha}\right)\right] ds,
\end{aligned} \tag{6.10}
$$

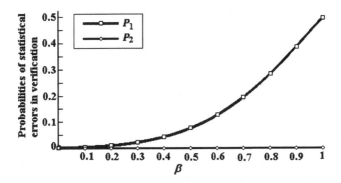

Fig. 16 Probability curve of statistical verification errors versus $\beta = m/\Delta$ – ratio of the systematic measurement error m to the limit Δ of the allowable error of MI being verified

where $k = \Delta/\sigma$, $\alpha = u/\sigma$, $\mu = m/\sigma = (m/\Delta)k$, $\beta = \mu/k = m/\Delta$.

Substituting into formulas (6.9) and (6.10) the known parameters: the characteristic k of serial suitability[2] of the type of MIs subject to verification and the standard uncertainty u of measurements at this standard, evaluated as type A, one may estimate the probabilities of verification errors for various values of systematic error m of measurements. Further, proceeding from the established bounds P_1 and P_2 of allowable values, the critical (maximum allowable) value of systematic error m_∂ of measurements is determined. The value m_∂ is a guide to metrological servicing of this standard, such as when setting up inter-calibration intervals, developing a method of statistical regulation of accuracy, and so forth.

Example 6.1. A verification unit has the following accuracy characteristics: $k = \Delta/\sigma = 3$, $\alpha = u/\sigma = 1/3$. The task is to determine the bound m^* of systematic measurement error so as to provide that the mean percentage of erroneous recognition of suitable MIs as unsuitable is no more than 5%.

Substituting $k = 3$ and $\alpha = 1/3$ into (6.9) and (6.10), we find the curves $P_1(\beta)$ and $P_2(\beta)$. They are presented in Figure 16. The figure shows that for these values of k and α, increasing m does not lead to increase in $P_2(\beta)$ (the maximum value of this probability is 0.00135). At the same time, $P_1(\beta)$ is very significant, and for $m = \Delta$ is 0.5. Substituting $P_1(\beta) = 0.05$, we obtain $\beta = 0.4$ and $m^* = 0.4\Delta$. Hence the systematic measurement error must not exceed 40% of the limit of allowable error of the MI being verified.

2. In developing standardized documents for state measurement chains and methods of verifying, it is essential to base the requirements for measurement accuracy,

---

[2] The characteristic of serial suitability of items, which is equal to the ratio of the allowance for the control parameter to the standard distribution of the values of this parameter throughout the set of items of this type, defines the probability of recognizing as suitable items of this type when under measurement monitoring. For successful passing this monitoring, it must be no less than three.

Table 23 Probabilities of statistical verification errors

α	K	P_1	P_2	K	P_1	P_2
$\frac{1}{1}$	2	0.128	0.0170	3	0.032	0.0011
$\frac{1}{1,5}$	2	0.065	0.0140	3	0.011	0.0010
$\frac{1}{2}$	2	0.0410	0.0120	3	0.0054	0.0008
$\frac{1}{3}$	2	0.0220	0.0100	3	0.0024	0.0007
$\frac{1}{5}$	2	0.0110	0.0070	3	0.0010	0.0005

in verifying any samples of MIs of one accuracy class, on any standard of the specified type. Here it is natural to assume that the error distribution of the standards of this type is subject to the normal law with zero mean and standard deviation u characterizing the dispersal of error among the set of standards of the indicated type. In this case, the probabilities of statistical verification errors are defined by formulas (6.9) and (6.10), in which $\beta = 0$, i.e. by the expressions

$$P_1 = \int_{-k}^{k} \varphi(s)\left[\Phi\left(-\frac{k+s}{\alpha}\right)+1-\Phi\left(\frac{k-s}{\alpha}\right)\right]ds,$$

$$P_2 = \int_{-\infty}^{-k} \varphi(s)\left[\Phi\left(\frac{k-s}{\alpha}\right)-\Phi\left(-\frac{k+s}{\alpha}\right)\right]ds + \int_{k}^{\infty} \varphi(s)\left[\Phi\left(\frac{k-s}{\alpha}\right)-\Phi\left(-\frac{k+s}{\alpha}\right)\right]ds.$$

$$(6.11)$$

Table 23 presents the values of these probabilities, calculated for a series of values α and k using formula (6.11).

The table shows that in order to ensure a percentage of verification errors at the level of 1–2%, it is necessary to take $\alpha \leq 1/3$. Proceeding from this, a widespread rule in metrology is: the limit of allowable error of an MI being verified must be at least 3 times larger than the limit of allowable error of the standard. This also involves the mechanism of improving the uncertainty of a transferred unit size when verifying an MI.

Hence, the overall framework for improving the uncertainty in the system of reproduction and transfer of unit size is presented as follows.

At the first l steps of the system, counting from the primary standard, the transfer is done by the calibration method. Increase in uncertainty is described by formula (6.5). In connection with this, the standard uncertainty of the standard at the l-th step is expressed by the formula

$$u_l = \sqrt{u_1^2 + \sum_{k=1}^{l-1}[u_{Ak}^2 + \sum_{i=1}^{n_k} u_{Bki}^2]}, \qquad (6.12)$$

where u_1 is the standard uncertainty of the unit size reproduced by the primary standard,

u_{Ak} is the standard uncertainty of measurements when transferring unit size from the standard at the k-th step, evaluated as type A,

u_{Bki} is the standard uncertainty of measurements when transferring unit size from the standard at the k-th step, caused by the i-th source of error, and evaluated as type B.

Further transfer of unit size proceeds by the verification method. Hence the standard uncertainty u_m of unit size preserved by the standard at the m-th step, is calculated by the formula

$$u_m = \frac{u_l}{\alpha_l \alpha_{l+1} \ldots \alpha_{m-1}}, \tag{6.13}$$

where $\alpha_l, \ldots \alpha_{m-1}$ is the ratio of standard uncertainties of unit sizes preserved by l standards, at the $(m-1)$-th step, to those preserved by standards (working MIs) being verified by them.

6.6 Measurement Chains

Documents that establish the order of transfer of unit sizes are called measurement chains. A measurement chain was first proposed in the 1930s by Professor L.V. Zalutsky for linear measurements. Thereafter they began to be developed to describe the coordination of reference standards and reference MIs in all types of measurement. Measurement chains are created both in other countries and by international metrological organizations.[3] In Russia, the content and construction of measurement chains is regulated by a state standard [37]. State and local measurement chains differ. A state measurement chain is foundational. It is distributed to all MIs of a given quantity that are verified and calibrated in the country. For MIs subject to verification, this document establishes the essential metrological requirements for instruments and the methods of verifying them. For MIs subject to calibration, these requirements are recommended. Local measurement chains, compiled on the basis of state measurement chains, are extended to MIs used in a separate department or in an enterprise. Local measurement chains define concretely the regulations of the state measurement chain by taking account of the specifics of their own measurements. Their development should be guided by the following principle: the accuracy of verification in compliance with a local measurement chain must be no less than that in a state measurement chain. This means, for example, verification of each MI using a standard of the same or higher area of the measurement chain; establishing that the ratio α of the bounds of allowable error of the standard to that of the MI being verified are no greater than the ratios indicated in the state measurement chain, and so forth.

---

[3] They are called calibrations chains in other countries and in international practice.

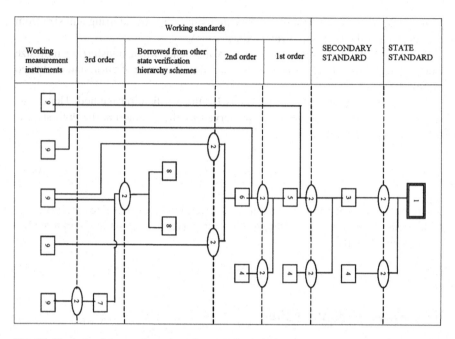

Fig. 17 Example of the configuration of a state (interstate) verification hierarchy schemes. 1 – state standard; 2 – verification method; 3 – secondary standard; 4 – transfer standard; 5–7 – working standards for corresponding orders; 8 – working standards borrowed from other state verification hierarchy schemes; and 9 – working MI

Measurement chains are formulated as a diagram (state measurement chains also contain a textual part, the explanations of the diagram). Figure 17 presents an example of a diagram of a state measurement chain. It consists of several horizontal bands, called the measurement chain fields. Each field corresponds to one level of transfer of the unit size. The following fields are distinguished: state reference standard, secondary standards, working standards (standard reference MIs) of the corresponding order (a separate field for each order), standards (standard reference MIs) borrowed from other state measurement chains, and working MIs. Rectangles are placed in the fields to show the names of the groups of MIs, the nominal values or ranges of values of the quantity and its error characteristics. These rectangles are joined by vertical lines corresponding to the recommended order of transfer of the unit size.

On the broken lines it is permissible to place ovals indicating the method of transferring the unit size, the error using this method, and additional information. Five methods of transfer are differentiated:

- *direct comparison,* used when comparing measuring instruments;
- *comparison with the aid of a transfer standard,* which is used in comparing measuring instruments that it is impossible to directly compare with each other;
- *comparison with the aid of a comparator,* used when comparing measures;

- *direct measurements*, used in verification and calibration of measures with regard to a reference standard measuring instrument, or of a measuring instrument with regard to a set of standard measures;
- *indirect measurements* (reproduction of a unit by the method of indirect measurements).

It follows from this description that information on the method of transferring the unit size is redundant, since the method is uniquely determined by the type of groups of MIs indicated in the rectangles. Hence, it is possible not to show but to collapse the ovals when there is no need for supplemental information.

It was shown in the preceding section that the uncertainty of the transferred unit size increases at all steps of the transfer of the unit size using the stages of the state measurement chain, from the state standard to the working MIs. Each superfluous step in the transfer of the unit size degrades the quality of verification and calibration of the working MIs at the end of this chain. Hence the number of fields of the state measurement chain must be the minimum necessary, fulfilling the condition that there be complete satisfaction of the need for verification and calibration of the working MIs used in the country.

Example 6.2. State measurement chain for an MI for the component percentage in gaseous media.

The state measurement chain for an MI for the component percentage in gaseous media is regulated by GOST 8.578–2008 [38]. It establishes the order and basic requirements for the transfer of the dimensions of the units of molar fraction (%) and mass concentration (mg/m^3) from the state primary standard with the aid of working standards, to the working MIs for component percentage in gaseous media. The state primary standard of units of molar fraction and mass concentration of components in gaseous media is at the head of this chain, and this standard includes unit assemblies of analytical and gas-mixing apparatus, transfer standards, and a set of specialized vessels. The standards units of equipment include:

- analytical unit for reproducing the unit of molar fraction of components in ranges from 2×10^{-8} to $5 \times 10^{-5}\%$ (for components of mixtures) and from 99.5 to 99.99995% (for a basic component);
- a gas-mixture gravimetric unit for reproducing the unit of molar fraction of inert, permanent, and chemically active gases, hydrocarbon components (methane, ethane, and others including liquefied hydrocarbons) and Freon, at intermediate scale points in a range from 1×10^{-4} to 99.5%;
- a gravimetric unit for reproducing the unit of mass concentration of chemically active gases and hydrocarbons in a range of yields from 2×10^{-2} to 50.0 μg/min;
- a unit of dynamic and volumetric large-scale conversion for reproduction of the unit of molar fraction of components in the range from 5×10^{-7} to 5% and the units of mass concentration of components in a range from 8.0×10^{-3} to 1.5×10^3 mg/m^3;
- a unit for reproduction and transfer of the unit size of molar fraction of ozone in a range from 5×10^{-8} to $1 \times 10^{-3}\%$;

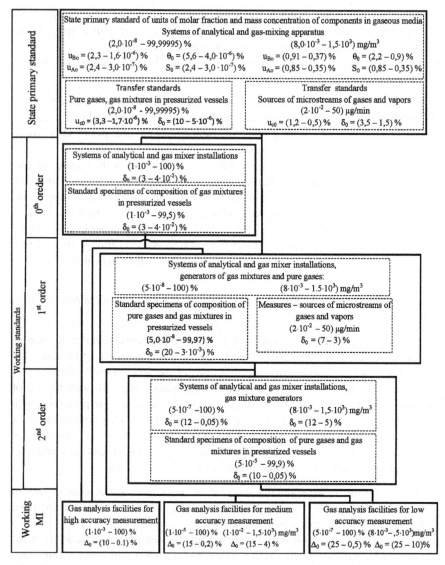

Fig. 18 State verification hierarchy scheme for substance ratio measurements in gaseous media (GOST 8.578-2008)

- analytical units for attestation of the transfer standards – pure gases and gas mixtures in pressurized vessels, in a range from 2×10^{-8} to 99.99995%,
- analytical units for attestation of the transfer standards – sources of microstreams of gases and vapors in a range of mass concentrations from 8.0×10^{-3} to 1.5×10^{3} mg/m^3.

Pure gases and gas mixtures in pressurized vessels are used as transfer standards in a range of molar fraction from 2×10^{-8} to 99.99995%, and sources of microstreams of gases and vapors in a range of yields from 0.02 to 50 μg/min.

The state primary standard is used for transfer of the dimensions of a unit of molar fraction and mass concentration of components to working standards of the 0th and first orders, and to working high-accuracy MIs by the methods of direct measurement and comparison using a comparator or transfer standards.

Units of analytical and gas mixer installations and reference standards of the composition of gas mixtures in pressurized vessels are used as 0th-order working standards. They are used to transfer the unit size of the molar fraction to first-order working standards and working high-accuracy MIs using methods of direct measurement and comparison with the aid of a comparator or standard references.

Units of analytical and gas mixer installations, generators of gas mixtures and pure gases, standard specimens of the composition of pure gases and gas mixtures in pressurized vessels, and sources of microstreams of gases and vapors are used as first-order working standards. They are used to transfer the unit size of the molar fraction to medium-accuracy and low-accuracy working standards, using methods of direct measurement and comparison with the aid of a comparator or standard references; and to transfer units of mass concentration to second-order working standards by indirect measurement methods.

Units of analytical and gas mixer installations, generators of gas mixtures and pure gases, and standard specimens of the composition of pure gases and gas mixtures in pressurized vessels are used as second-order working standards. They are used to transfer the unit sizes of the molar fraction and mass concentration to low-accuracy working MIs, using methods of direct measurement and comparison with the aid of standard specimens of the composition of pure gases and gas mixtures of the second-order in pressurized vessels.

Specialized and universal gas analyzer MIs (gas analyzers, gas analysis stations and posts for monitoring atmospheric pollution, mass spectrometers, and other types of gas analysis equipment) are used as working MIs. These pertain to one of three levels of accuracy: high, medium, and low. Figure 18 shows the metrological characteristics of the standards and working MIs.

6.7 Organizational Basis of Measurement Assurance

Management of a system of measurement assurance in a country is a most important governmental function. Analysis of the experience of foreign countries has shown that even here this activity is in the sphere of state management. In the USA, this function is assigned to the Department of Commerce, under which is the National Institute of Standards and Technology (NIST). In England, it is the Secretary of State for Trade and Industry, which conducts the National Physical Laboratory, and in Germany, it is the Department of Economics, to which the Physico-technical

Institute (PTB) is subordinated [39]. Since May 20, 2004, this function has been performed in Russia by the Federal Agency for Technical Regulation and Metrology (Rostekhregulirovaniye), which is within the system of federal executive agencies of the Russian Federation and is located under the RF Ministry of Industry and Energy.[4] In the realm of metrology, Rostekhregulirovaniye executes the functions of state metrology monitoring and supervision. The RF law "On Measurement Assurance," adopted in 1993, established the following types of state metrology control:

- approval of the type of MI,
- verification of MIs, including reference standards,
- licensing the activity of legal entities and persons to the right to manufacture, repair, sell, and lease MIs.
- The types of state metrological supervision are:

  - supervision over the condition and use of MIs, methods of performing measurements, and observation of other metrological rules and regulations,
  - supervision over the quantity of packaged goods in any type of packing at their unpacking, sale, and import,
  - supervision over the quantity of goods disposed of at the completion of trade operations, as well as regarding the condition and usage of gaming machines with monetary payout, and their software.

Rostekhregulirovaniye organizes and coordinates activity on measurement assurance in Russia, including activity subordinate metrological organizations, which include:

- Eight scientific research institutes [NIIs] that perform the functions of Russia's NMIs – D.I. Mendeleev VNIIM, VNIIMS, VNIIFTRI, VNIIOFI, VNIIR, VS VNIIFTRI, VNIIFI Dal'standart, and VNIITsSMV;
- State time and frequency service and determination of the parameters of the Earth's rotation;
- State service for standard reference data on physical constants and the properties of substances and materials;
- State service for standard specimens of the composition and properties of substances and materials;
- 86 Centers for Standardization, Metrology, and Certification (SMS) which perform the functions of state metrological control (verification of MIs and other types of metrological monitoring enumerated above);

---

[4] Beginning in 1954 when the Committee on Standards, Measures and Measuring Instruments under SM SSSR was first organized on the basis of the Guidance on Standardization under Gosplan SSSR and the D.I. Mendeleev VNIIM, the agency of state administration of metrology, with different names but analogous functions, has always been within the federal executive agencies.

– territorial agencies of state supervision in all autonomous republics, regions, and oblasts of the country, which perform the functions of state metrological control.

In addition, Rostekhregulirovaniye participates in the activity of international organizations on metrology, directly and through the subordinate NMIs, and is an official representative of the Russian Federation in these organizations.

The basic task of Russia's NMIs is the conduct of fundamental research in the area of metrology; creation, improvement, preservation, and use of state standards; and participation in international work in the area of metrology. The NMIs reproduce on their own state standards the units of physical quantities and transfer their dimensions to the SMSs; i.e., to secondary and working standards that are the reference standards of these organizations. Further, the SMSs transfer the unit sizes to the reference standards of metrological services of executive agencies and legal entities, with the help of which working MIs are verified and calibrated. Hence an organizational system for the reproduction of a unit and transfer of its size has been set up, analogous to the technical system studied in Sect. 3. Figure 18 shows the block diagram of this system.

6.8 Measurement Assurance in the World on the Basis of the Arrangement Between the Directors of the National Metrology Institutes

After the Metric Convention was signed in 1875, the degree of uniformity of measurement on a planetary scale gradually increased. This activity is headed by the CGPM and its agencies, a diagram of which is shown in Figure 19. The CIPM and BIPM coordinate work in the area of uniformity of measurement of regional metrological organizations (RMOs), a list of which is presented in Table 24.

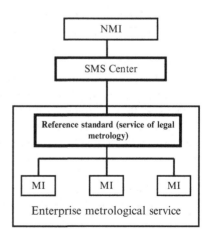

Fig. 19 Organizational system for reproducing a unit and transferring its size

Table 24 Regional metrological organizations

No.	Designation	Notation	
		English	Russian
1.	Asian-Pacific Metrological Program	APMP	АПМП
2.	Euro-Asian Cooperation of National Metrology Institutions	COOMET	КООМЕТ
3.	European Association of National Metrological Institutes	EURAMET	ЕВРАМЕТ
4.	Inter-American Metrology System	SIM	СИМ
5.	South African Development Community Cooperation in Measurement Traceability	SADCMET	САДКМЕТ

In recent decades, this process has proceeded at significantly faster tempos. This is explained by the immense changes that accumulated toward the end of the second millennium and are characterized by the term "globalization." Above all, this pertains to the development of international productive cooperation and trade based on international standardization, international unification of transport systems, elimination of all barriers between national economies, leading to their actual unification into a unified world economic system. This is also the creation of a unified information space and continually broadening international cooperation in various areas – science, education, medicine, ecology, meteorology, agriculture, and other. As a consequence of globalization there has arisen the problem of the recognition, by all countries of the world, of the results of measurements and tests conducted in each country. For example, trade agreements between states require this from all parties signing a contract. Uniformity of measurements is also a necessary condition for successful performance of international scientific programs and the solution of many other issues. Thus, at the present time, the BIPM is initiating the task of creating a worldwide system to ensure uniformity of clinical medical analyses [40].

The resolution of this issue is linked to the creation of a system for accreditation of measurement and test laboratories. But all MIs of these laboratories are subject to calibration or verification. Hence it is necessary also to create a system for accreditation of calibration and verification laboratories, and this provides a mechanism for international recognition of certificates of calibration and verification. And finally, since these laboratories receive the unit sizes from national measurement standards kept at NMIs, coordination of the dimensions of these units is required.

The necessity for coordination comes from the fact that centralized systems of measurement assurance are limited by the scope of the separate states. As a result, the dimensions of a unit reproduced or preserved by the national standards of different countries may significantly differ from each other. Figure 20 shows a framework illustrating the described cause and effect relationships.

The traditional method of conformity consists of creating a primary standard, from which the unit size is transferred to all national standards. However, on a planetary scale, this is a very expensive and technically hardly feasible path to achieve the stated goal. Hence an international system currently being actively developed for confirming the metrological equivalence of national standards is resolving this issue (*"metrological equivalence" of standards is understood to be*

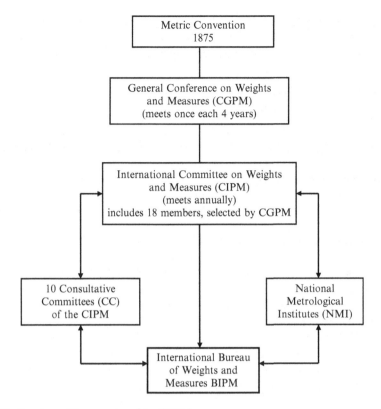

Fig. 20   Structure of the agencies of the CGPM

the correspondence of the degrees of accuracy of the units reproduced and (or) preserved by them).

The beginning of the system of international recognition of national standards was set in place by CIPM, which on October 14, 1999 opened up for signing by the directors of the NMIs a document entitled "Mutual Recognition Arrangement [MRA]. Mutual Recognition of the National Measurement Standard and Calibration and Measurement Certificates Issued by the National Metrology Institutes" (hereinafter "DVP") [41]. In this, the following goals were proclaimed:

- establishment of the degree of metrological equivalence of national standards,
- ensuring mutual recognition of calibration and measurement certificates issued by the NMIs,
- ensuring a reliable technical base for broader arrangements in the area of international trade, commercial activity, and standardization; in particular, arrangements on mutual recognition of calibration, measurement, and testing certificates issued by accredited laboratories of other countries.

Key comparisons produced by the CIPM and RMO Consultative Committees for types of measurements were set as the foundation of this system. In addition, the DVPs require also the fulfillment of the following conditions:

- a functioning quality control system in each NMI and confirmation of the competence of the NMI,
- successful participation of each NMI in subsequent comparisons.

This process is concluded by the publication of announcements regarding the measuring capabilities of each NMI, to be incorporated into the BIPM database and published on the Internet at the BIPM website.

Key comparisons consist of the measurement by all participants of the value of a quantity preserved by a standard of comparison or other highly stable measure, and then comparing the results of these measurements. Key comparisons for the CIPM are done by the Consultative Committees (CCs) of types of measurements. The CC at each session studies the need for comparisons and decides, taking account of the opinions of the RMOs, which comparisons need to be initiated. Laboratories (NMIs) that participate in the CCs, and other laboratories that have the highest technical competence and experience in the given type of measurement, are invited to participate in the key comparisons by the CIPM. To organize and conduct the comparison, a pilot laboratory and 2–3 laboratories to assist the pilot laboratory are designated. The pilot laboratory, together with the designated assisting ones, compiles a detailed technical protocol (a technical reference manual that describes in detail the procedure for conducting the comparison) and a schedule for the comparisons. Then the pilot laboratory prepares and studies a standard of comparison and organizes its shipment to the participants according to the schedule of comparisons. The laboratories that are participants in the comparison will, independently of each other and in strict compliance with the technical protocol, conduct measurements of the measure that was sent. Based on these measurements, they determine the value of the measure and its uncertainty. Then they present these results with the necessary supplemental information to the pilot laboratory. The pilot laboratory analyzes and processes the obtained results and determines the reference value and its extended uncertainty [42].

The reference value is understood to be the value of the quantity, the measurement uncertainty of which is recognized by all to be sufficiently insignificant that it may serve as a basis of comparison for values of quantities of the same type. It follows from this definition that the reference value of a quantity is an analogue of the true value of a quantity. For example, in measuring mass, where the mass of an international prototype of the kilogram is by definition the unit, the reference value of the unit is the value assigned to this prototype. In other types of measurements, there is no such prototype, and hence what is accepted as the reference value of a unit is the mean weighted arithmetic average of a number of its values reproduced by the most accurate national standards that are preserved by the NMI participants in the key comparison for the CIPM:

$$x_{\text{CIPM}} = \sum_{i=1}^{n} x_i p_i, \tag{6.14}$$

where x_i is the result of measuring the value of the standard of comparison at the ith national standard,

$$p_i = \frac{1/u_i^2}{\sum_{i=1}^{n} 1/u_i^2}$$

is the "weight" of the measurement result x_i, u_i is the standard uncertainty of this measurement result, n is the number of NMIs participating in the comparison.

The standard uncertainty of the reference value with independent values x_i is equal to

$$u(x_{\text{CIPM}}) = \frac{1}{\sum_{i=1}^{n} 1/u_i^2}. \tag{6.15}$$

The CIPM key comparison reference value is the initial one in the system of comparisons. It is taken as a reference point in determining the RMO key comparison reference values, which in turn fulfill the function of a reference point in conducting supplemental comparisons. Overall, a hierarchical chain of comparisons of national standards has been built up, being an analogue of the hierarchical chain "primary standard – secondary standard – working standard". In essence, the idea of a centralized system of measurement assurance in the world has been implemented in the system of comparing national standards.

The CIPM key comparison reference value is the initial value not just in the system of comparisons, but also generally in the worldwide system of ensuring measurement uniformity of a given type. Through the system of comparisons, this value is transferred to national standards and from them, through national systems for transferring dimensions of units, to all MIs that are calibrated in this unit (see Fig. 21). If one represents that the national standards participating in the CIPM key comparisons are a unified collective standard confirmed as a worldwide primary standard, then the reference value will be the value of a unit reproduced by this world standard.

But the CIPM key comparison reference value has also another important function: it serves to evaluate the degree of equivalence of national standards. By definition, *the degree of equivalence of a national standard is that degree to which the value of a standard corresponds to the CIPM key comparison reference value* [11]. Hence it is expressed quantitatively as a deviation from the reference value

$$d_i = x_i - x_{\text{CIPM}} \tag{6.16}$$

in conjunction with the standard uncertainty of this deviation, calculated by the formula

$$u(d_i) = \sqrt{u^2(x_i) + u^2(x_{\text{CIPM}}) - 2\text{cov}(x_i, x_{\text{CIPM}})}, \tag{6.17}$$

where $\text{cov}(x_i,\ x_{\text{CIPM}})$ is the covariance of the value x_i and the reference value.

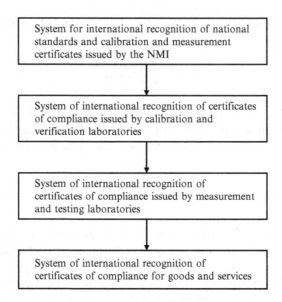

Fig. 21 Hierarchical structure of the systems for international recognition of certificates of conformity

The claimed accuracy of a national standard is confirmed with probability 0.95 if this inequality is satisfied:

$$|d_i| < 2u(d_i), \tag{6.18}$$

If this condition is not fulfilled for any national standard, that would provide evidence that the accuracy claimed by the NMI was not confirmed by comparison. Consequently, the obligatory condition of international recognition of calibration and measurement certificates, issued to the given NMI, would turn out to be unfulfilled. This circumstance can lead to serious consequences for the NMI and the country that it represents since, as seen in Fig. 20, a refusal to recognize calibration and measurement certificates in some type of measurement that were issued by the NMI will automatically lead to the legal invalidity of the certificates of conformity of all measurements results of that type that were conducted in that country. In order not to reach that point, there are frequently inspections directed toward increasing the estimate of uncertainty of this standard, and information about its measuring capabilities is published by the BIPM with its enhanced estimate of uncertainty.

RMO key comparisons are conducted using a similar setup. Their goal is the dissemination of metrological equivalence to the national standards of the RMO countries that do not take part in the CIPM key comparisons. All national standards of member-nations of the RMO that did not participate in the CIPM key comparisons are invited to take part in these key comparisons. Besides these,

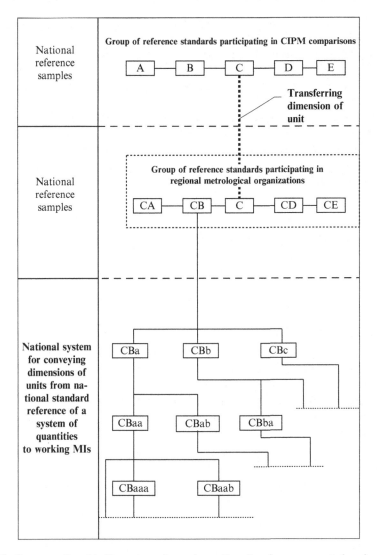

Fig. 22 Structure of worldwide system of ensuring uniformity of measurements based on Key Comparisons

one or several national standards that have participated in the CIPM key comparisons are invited. These are called binding standards. The binding national standards in the RMO key comparisons do not confirm their own equivalence, since this was done in the CIPM key comparisons. They fulfill the function of transferring the CIPM's key comparison reference value to the RMO.

Director of TK COOMET "Key Comparisons" A.G. Chukovkina [43, 44] developed the methodology for conducting RMO key comparisons and statistical analysis of their results. These are done as follows. After the formation of a group of participants and the design of a technical protocol and schedule of comparisons, the pilot laboratory sends the standard of comparison to the participants in the comparisons. All participants conduct measurements of the value of the standard of comparison and derive the measurement results $\tilde{x}_i$ and an estimate of their standard uncertainty $u(\tilde{x}_i)$.

After completion of the measurements, it is necessary to reduce their results $\tilde{x}_i$ to the CIPM's key comparisons reference value. With that goal, the pilot laboratory does the following.

1. Based on the measurement results of the binding standards, it calculates the additive correction Δ to the measurement results $\tilde{x}_i$ and the estimate of standard uncertainty $u(\Delta)$ of this correction:

$$\Delta = \frac{\sum_{i=1}^{l} \Delta_i/S_i^2}{\sum_{i=1}^{l} 1/S_i^2}, \tag{6.19}$$

$$u(\Delta) = \sqrt{\frac{2}{\sum_{i=1}^{l} 1/S_i^2}}, \tag{6.20}$$

where $\Delta_i = x_i^* - \tilde{x}_i^*$,

x_i^*, $\tilde{x}_i^*$ are the measurement results of the binding standard in the CIPM and RMO key comparisons, respectively,

l is the number of binding standards;

2. It determines the *transformed (reduced to the CIPM key comparisons reference value)* measurement results of all laboratories that are participating, and their uncertainties, using the formulas

$$\tilde{x}_i^\circ = \tilde{x}_i + \Delta, \quad i = 1, ..., n, \tag{6.21}$$

$$u(\tilde{x}_i^\circ) = \sqrt{u^2(\tilde{x}_i) + u^2(\Delta)}, \tag{6.22}$$

where n is the number of standards participating in comparisons.

Further, the degree of equivalence is calculated for all national standards participating in the comparison other than the binding standards

$$d_i = \tilde{x}_i + \Delta - x_{\text{CIPM}} \tag{6.23}$$

and its standard uncertainty

$$u(d_i) = \sqrt{u^2(x_i) + u^2(x_{\text{CIPM}}) + u^2(\Delta)[1 - u^2(x_{\text{CIPM}}) \sum_{j=1}^{l} \frac{1}{u^2(x_i^*)}]}. \tag{6.24}$$

Then they verify the equivalence of these standards. If inequality (6.18) is fulfilled, then the estimates of uncertainty of results, claimed by the NMI, agree with a probability of 0.95 with the data of the comparisons. This is verification of the claimed accuracy of the national standard. In this case, the claim of the measuring capabilities of the NMI is entered into the BIPM database and published on the Internet at the BIPM website. If inequality (6.18) is not fulfilled, then this provides evidence that the accuracy claimed by the NMI for the national standard was not confirmed. Then they reexamine on the side of increasing the estimate of uncertainty of this standard, and information about its measuring capabilities is published with its enhanced estimate of uncertainty.

They determine the actual values of the units $\hat{x}_i$, kept by the national standards, and the estimates $u(\hat{x}_i)$ of their uncertainty:

$$\hat{x}_i = Q_{RMO}(\tilde{x}_i - d_i), \tag{6.25}$$

$$u(\hat{x}_i) = \frac{\partial Q_{RMO}}{\partial x} \sqrt{u^2(\tilde{x}_i) + u^2(d_i) - 2\mathrm{cov}(\tilde{x}_i, d_i)}, \tag{6.26}$$

where Q_{RMO} is a known function of the link between the value of the unit and the value of the standard of comparison used in the MRO key comparisons.

Further, they estimate the RMO key comparisons reference value and its standard uncertainty:

$$x_{RMO} = \sum_{i=1}^{n} \frac{\tilde{x}_i/u^2(\tilde{x}_i)}{\sum_{i=1}^{n} 1/u^2(\tilde{x}_i)} + \Delta, \tag{6.27}$$

$$u(x_{RMO}) = \sqrt{\frac{1}{\sum_{i=1}^{n} 1/u^2(\tilde{x}_i)} + u^2(\Delta)}. \tag{6.28}$$

There is a significant difference between the CIPM and RMO key comparisons reference values: the reference value x_{RMO} does not reflect an actual value of a unit transferred to the national standards of the countries of this region, and hence is not involved in the system of transferring the unit size. Its function is different: x_{RMO} serves to define the mean degree of equivalence of the group of these standards. It is equal to

$$d_{RMO} = x_{RMO} - x_{CIPM}, \tag{6.29}$$

and the estimate of its uncertainty is

$$u(d_{RMO}) = \sqrt{u^2(x_{RMO}) + u^2(x_{CIPM}) - 2\mathrm{cov}(x_{RMO}, x_{CIPM})}. \tag{6.30}$$

d_{RMO} and $u(d_{RMO})$ are generalized characteristics of the accuracy of national standards and high-accuracy measurements in this region. They can serve as criteria for the relative evaluation of different RMOs as to the degree of accuracy of measurements in the NMIs.

At present, the system for confirming the measuring capabilities of NMIs using key comparisons of national standards is being actively developed. Hence by the beginning of 2007, there were 356 of just the CIPM key comparisons registered in the BIPM database [45]. The positive effect of these comparisons on the degree of uniformity of measurements has been noticed by all specialists. It has become evident that opening up the DVP for signing by the NMIS was the greatest step in ensuring uniformity of measurement in the world, comparable in significance to the signing of the Metric Convention.

Chapter 7
Verification and Calibration Intervals of Measuring Instruments

7.1 Basic Concepts

It was shown in Sect. 6.2 that among the basic elements of a system for metrological assurance are periodic verifications and calibrations of MIs, in which there is monitoring of the compliance of metrological characteristics of MIs to the established requirements, or of their behavior in compliance with these requirements. The verification or calibration interval (VCI) is the most important parameter of metrological maintenance of the MI, directly affecting the level of uniformity of measurement. The lower the VCI, the higher this level is. On the other hand, the lower the VCI, the greater the financial expenses to conduct verifications and (or) calibrations of MIs, as well as production cost associated with removing the MI from the place of use. Hence there is an opposition, which must be resolved by determining the optimal value of the VCI.

The most natural criterion for optimality of a VCI is the economic criterion – the conditional minimum of the economic costs of using the MI, which accumulate from losses due to insufficient accuracy in measurements and of expenses associated with performing verifications, calibration, and repairs of MIs that were rejected in verification.

Figure 23 presents a graphic illustration of this approach to determine the calibration interval. It is clear that the mean annual total costs incurred for one MI, accumulated due to losses from measurement defects and recalibration expenses, are lowest in the range [0.6–1.2] years. Consequently, it is advisable to take VCI as 1 year.

Methods for substantiating a VCI were developed within the scientific specialization of theoretical metrology, which has received the designation of "theory of metrological reliability of measuring instrument" [46–48].

The basic concepts and definitions of theoretical metrology are regulated in [34, 49, 50]. Their interrelationship is illustrated in Fig. 24, which shows the

A.E. Fridman, *The Quality of Measurements: A Metrological Reference*, DOI 10.1007/978-1-4614-1478-0_7, © Springer Science+Business Media, LLC 2012

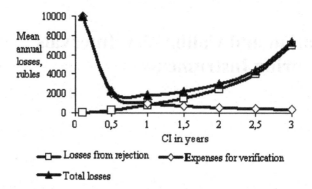

Fig. 23   Determining the optimal VCI using economic criteria

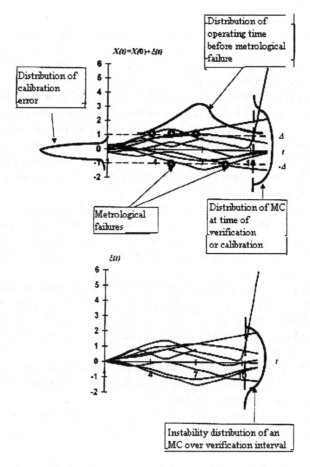

Fig. 24   Change in metrological characteristics of an MI over time

trajectories of the random process of changing the metrological characteristics (MC) of one type of MI (such as the systematic component of base error, the SD of the random component of base error, or other MC defined in accordance with [17]).

The intersection of the trajectories of the bounds "$\pm\Delta$" of the allowable values of the MC is a metrological failure and a break in the trajectory is a failure of functioning. The indicators of accuracy, stability, and metrological reliability of the MI correspond to various functionals constructed on these trajectories. Accuracy of the MI is evaluated by the value of the MC at an instant in time being examined, and for a set of MIs, by the distribution of these values (the MC distributions after initial calibration and the time of verification are shown). *The stability of an MI, reflecting the invariability of its properties in time,* is estimated by the distribution of the increments of the MC over the specified time. *Metrological reliability of the MI is defined as reliability of the MI in regard to preservation of metrological serviceability; i.e., the state in which all MCs correspond to the established norms.* Consequently, it is evaluated by the distribution of times that metrological failures occurred. It is clear that the metrological reliability of the MI depends on its stability, but in addition on the established guidelines and conditions of operation of the MI – the bounds of the allowable values of the MC, calibration method, VCI, and others. Hence by comparison with stability, it is as though it were an "external" property of the MI. From this ensues the following statement of principle: the evaluation of metrological reliability of an MI is possible only after examining its stability.

Below are the basic terms of the theory of metrological reliability, with their definitions.

1. *Metrological serviceability of an MI – the state of an MI, determined by the compliance of its normalized MCs with established requirements.*
2. *Metrological reliability of an MI – reliability of an MI in regard to preserving its metrological serviceability.*
3. *Metrological failure of an MI – failure of an MI consisting of the loss of metrological serviceability.*
4. *Stability of an MI – a qualitative characteristic of an MI, reflecting the invariability of its MC over time.*
5. *Instability of an MI's MC – change in an MI's MC over a set interval of time.*

Note: If the stability of an MI is evaluated on one of its MCs, the term "*instability of the MI*" may be used instead of this term.

6. *Mean instability of an MI's MC – an indicator of the instability of an MI's MC, equal to the mean of the instability of this characteristic, over a group of MIs of one type or over a set of periods of use of one MI.*
7. *Standard deviation of instability of an MI's MC – an indicator of the instability of an MI's MC, reflecting the dispersal of the instability over a group of MIs of one type or over a set of periods of use of one MI, and equal to the square root of the dispersion.*

Note: The term "*standard uncertainty of instability of the MC*" may be used instead of this term.

8. *Confidence limits of the instability of an MI's MC – the upper and lower bounds of the interval that encompass the instability of an MI's MC with some confidence probability.*

Note: The term "*extended uncertainty of the instability of the MC*" may be used instead of this term.

9. *The probability of the metrological serviceability of an MI – an indicator of the metrological reliability of an MI, equal to the probability that at the specified instant the MI is metrologically serviceable.*
10. *The probability of operation of the MI without metrological failures (the probability of failure-free operation) – an indicator of the metrological reliability of an MI, equal to the probability that there is no metrological failure of the MI in the course of a specified running time or a specified operational time.*
11. *Metrological serviceability coefficient of an MI – an indicator of the metrological reliability of an MI, equal to the average percentage of the VCI in which the MI is in a metrologically serviceable state.*
12. *Verification (calibration) interval – the span of time or running time of an MI between two consecutive verifications (calibrations).*

7.2   Basic Concepts of the Theory of Metrological Reliability

7.2.1   Mathematical Model of Metrological Characteristics Drift

For simplicity of presentation, we shall examine the MCs of an MI that are characterized by one numerical value throughout the range of measurements. This includes the transform coefficient of a measuring transducer, the basic error of a single measure, the standard deviation of the random component of the base error of the MI, the characteristics of the additive and multiplicative errors with a linear transform function, its dynamic errors, and others. This limitation is not based on principle, since in [51] it was shown that the obtained results are easily extended to MCs, the values of which are used on the MI's scale.

Let us examine one of these MCs. In the set of MIs of one type, its values are random quantities with the same distribution, but their changes over time are random trajectories. Hence the mathematical model of the drift process of this MC is a non-stationary random process. The following hypothesis is assumed in order to make this model more concrete:

The trajectory of the drift of an MI's MC at any instant is continuously differentiable with probability 1.

This signifies that step-function changes in the rate of drift of the MC are of low probability.[1] In other words, almost every trajectory of the drift process at any instant t has a bounded first derivative $\xi'(t)$, the values of which on both left and right are equal to $(\xi_{-}'(t) = \xi_{+}'(t))$.

One example of this process is the non-random function of random arguments $\xi(t) = f(t, \alpha_1, ..., \alpha_n)$, where α_i $(i = 1, ..., n)$ are random quantities, and are generally correlated.

This kind of hypothesis corresponds fully to current descriptions of the processes of aging and wear and to the method of presenting the initial data on MC drift (mean instability $m(t)$ and standard deviation of instability $\sigma(t)$ are provided as smooth curves in time). In addition, this model accounts for the possibility of gathering the test material on the instability of the MI, since the values of MC are monitored through defined intervals of time, and consequently, it is not possible to record instantaneous jumps in the drift speed when conducting a test. An advantage of this model is also that it does not present a monotonic change in MC that does not correspond to actual concepts on the instability of many types of MIs.

7.2.2 Basic Instability Equation and Its Solution

Based on this hypothesis and using strict mathematical transformations, a differential equation with respect to the unknown function $P(t, x, y)$ has been derived, equal to the probability that the instability $\xi(t)$ of the MC over time t lies between arbitrary bounds x and y:

$$\frac{\partial P(t,x,y)}{\partial t} + \mu(t,x)\frac{\partial P(t,x,y)}{\partial x} + \mu(t,y)\frac{\partial P(t,x,y)}{\partial y} = 0, \qquad (7.1)$$

where t is the interval of time or running time of the MI from the beginning of operation until the examined instant,

$\mu(t, \xi)$ is the conditional expectation of the speed $v(t)$ of MC drift at time t under the condition that MC instability at time t equals $\xi(t) = \xi$.

In probability theory, the conditional expectation of quantity Y on the condition, that quantity X has a specific value, is called regression of Y on X. Hence $\mu(t, \xi)$ *is called regression of the instantaneous speed $v(t)$ of MC drift on instability $\xi(t)$ at time t.*

Equation (7.1) is called the fundamental equation of instability.

Considering that $\mu(t, \xi)$ is continuous on t and ξ, we shall find the solution of equation (7.1) at the natural initial conditions $x|_{t=0} = x, y|_{t=0} = y, P(t,x,y)|_{t=0} = P(0,x,y) = \int_x^y \varphi_0(\xi)\,d\xi$, where $\varphi_0(\xi)$ is the probability density ξ at the initial

---

[1] To understand the model, it is important to keep in mind that what is modeled is not the change of error measurements defined by the MI's MC, but the change in these MCs; measurement error, including the random component, naturally, can change by a step function at any instant.

instant of time. It is easy to show that this always exists, and is unique. The solution
is the function

$$P(t, x, y) = \int_x^y \varphi_0[\Psi(t, \xi)] \frac{\partial \Psi(t, \xi)}{\partial \xi} \, d\xi, \qquad (7.2)$$

where $\Psi(t, \xi)$ is the solution of the differential equation of regression

$$\frac{d\xi}{dt} = \mu(t, \xi) \qquad (7.3)$$

in the form $\Psi(t, \xi) = C$ with initial condition $\Psi(0, \xi) = \xi$ (hereafter, we shall call
this the regression function).

It follows from (7.2) that the probability distribution of MC instability at an
arbitrary instant t is equal to

$$\varphi_t(\xi) = \varphi_0[\Psi(t, \xi)] \frac{\partial \Psi(t, \xi)}{\partial \xi}. \qquad (7.4)$$

Hence, the probability distribution of MC instability at an arbitrary instant is
equal to its initial probability density, multiplied by an appropriate normalizing
multiplier, into which the regression function is substituted as a variable.

It may be proven that the regression function has the following properties:

1. $\Psi(t, \xi)$ is a monotonic function of ξ.
2. $\Psi_t(\tau, \xi)$ has the property of reproducibility:

$$\Psi_z[\tau, \Psi_{z+\tau}(t, \xi)] = \Psi_z(\tau + t, \xi).$$

3. $\frac{\partial \Psi(t,\xi)}{\partial t} = - \frac{\partial \Psi(t,\xi)}{\partial \xi} \mu(t, \xi)$.

One may also prove that $\varphi_0(\xi)$ is the normal probability density.[2] Formulas (7.3) and
(7.4) make it possible to substantiate the possible types of distribution laws of MC
instability.

7.2.3 Instability Distribution Laws

In technical appendices, it often turns out to be sufficient to approximate the
regression with a linear mean square function. A linear mean square regression of
the rate of MC drift on its change over time t is expressed by the function

---

[2] The development of equation (7.1), its solution, and the mathematically strict substantiation of
other results of this chapter are presented in Chapter 9 of the composite monograph "Theory of
Metrological Reliability of Measuring Equipment" [47].

$$\mu(t, \xi) = m'(t) + r_1(t)[\xi - m(t)], \tag{7.5}$$

where $m(t)$ is the mean of the distribution $\varphi_t(\xi)$,

$r_1(t) = \frac{\sigma'(t)}{\sigma(t)}$ is the coefficient of regression, and

$\sigma(t)$ is the standard deviation of the distribution $\varphi_t(\xi)$.

In this expression, the derivative nature of the MC drift is manifested in a simpler and more natural manner: at any instant the deviation of the rate of MC drift of a specific MI from the average rate of MC drift is directly proportional to the deviation of the value of the instability of the MC of this MI for the preceding period from the mean instability of MC for the same period.

Substituting the specified function into (7.3) and solving it, we derive the regression function

$$\Psi(t, \xi) = G(t, \xi)\sigma(0) + m(0). \tag{7.6}$$

where $G(t, \xi) = \frac{\xi - m(t)}{\sigma(t)}$ is the drift function.

With this, the initial form of the MC instability distribution is preserved, and only its mean $m(t)$ and SD $\sigma(t)$ are modified:

$$\varphi_t(\xi) = \varphi_0 \left[(\xi - m(t)) \frac{\sigma(0)}{\sigma(t)} + m(0) \right] \frac{\sigma(0)}{\sigma(t)}.$$

Hence, with a linear mean square regression of the rate of MC drift on its instability, the probability density of its instability is subject to the normal law:

$$\varphi_t(\xi) = \frac{1}{\sqrt{2\pi}\sigma(t)} \exp\left\{ -\frac{[\xi - m(t)]^2}{2\sigma^2(t)} \right\}. \tag{7.7}$$

More general concepts are based on accounting for the regularity of the course of the physical and chemical processes that cause aging and wear on the MI. In the first approximation, the equation for the rate $v(t)$ of any process of deterioration may be presented as a non-random function with random arguments

$$v(t) = s'(t) = Q[T, \alpha_1, \dots, \alpha_n]s^{1-F}(t), \tag{7.8}$$

where $Q[T, \alpha_1, \dots, \alpha_n]$ is the coefficient of proportionality depending on temperature T and n random parameters α_i, $i = 1, \dots, n$, that characterize the chemical composition and properties of materials, surface properties of materials, configuration of the item, the environment, loads imposed, and other similar factors,

$s(t)$ is the value of the defining parameter of the process at time t,

F is the level indicator, different for different processes, but having a defined value for each specific process.

Let us compare (7.8) with the expression for the linear regression of $v(t)$ on $s(t)$. For $F = 1$ $m'(t) = M[Q]$ and $m(t) = M[Q]t$, where $M[Q]$ is the mean of $Q[T, \alpha_1, ..., \alpha_n]$. For a specific trajectory, we have $\xi_i'(t) = Q_i$ and $\xi_i = Q_i t$. Hence $\xi_i'(t) = Q_i = M[Q] + (Q_i - M[Q]) = m'(t) + r_1(t)[\xi_i - m(t)]$, since

$$r_1(t) = \frac{\sigma'(t)}{\sigma(t)} = \frac{\sigma[Q]}{\sigma[Q]t} = \frac{1}{t}.$$

Hence, for $F = 1$ the linear mean square regression of $v(t)$ on $s(t)$ coincides with the precise function of this regression, and the approximation error $\mu(t, \xi)$ becomes equal to zero. From this, it follows:

the normal distribution law precisely describes instability in those cases when the mean rate of drift at any instant does not depend on the object's instability for the preceding period.

As a rule, rates of corrosion, linear surface erosion, and certain other processes of deterioration that occur from the effect of mechanical loads correspond to this condition [52]. In all other cases, the use of a model of linear mean square regression of $v(t)$ on $s(t)$ will lead to additional methodological error. However, this can be eliminated by simple transformations.

For example, a whole series of chemical processes of aging, such as recrystallization, diffusion, chemosorption, and other heterogeneous 1st-order reactions, as well as some processes of mechanical damage (such as development of cracks) are characterized by $F = 0$ [52]. For this case, let us divide the left and rights sides of equation (7.8) by s. Then the equation is written as:

$$[\ln |s(t)|]' = Q[T, \alpha_1, ..., \alpha_n].$$

Since the derivative of the logarithm of $|s(t)|$ depends only on Q, a linear model will be a precise description of its regression on $\ln|s(t)|$. Hence we obtain, by analogy with the preceding,

$$\mu(t, \ln |\xi|) = m'_{\ln}(t) + r_1(t)[\ln |\xi| - m_{\ln}(t)],$$

where $r_1(t) = \sigma'_{\ln}(t)/\sigma_{\ln}(t)$, and $m_{\ln}(t)$ and $\sigma_{\ln}(t)$ are the mean and standard deviation of the function $\ln|s(t)|$, defined by the formulas

$$m_{\ln}(t) = \int_{-\infty}^{\infty} \text{sign}\xi \times \ln |\xi| \times \varphi_t(\xi)\, d\xi,$$

$$\sigma_{\ln}(t) = \left\{ \int_{-\infty}^{\infty} \left(\text{sign}\xi \times \ln |\xi| - m_{\ln}(t) \right)^2 \times \varphi_t(\xi)\, d\xi \right\}^{0.5}.$$

From the obvious equality $\xi' = \xi(\ln|\xi|)'$ it follows that $\mu(t, \xi) = \xi\mu(t, \ln|\xi|)$. This signifies that for drift intensity in (7.1) and (7.3) is derived the expression

$$\mu(t, \xi) = \xi\{m'_{\ln}(t) + r_1(t)[\xi - m_{\ln}(t)]\}. \tag{7.9}$$

From this, for the initial condition $\Psi(0, \xi) = \text{sign}\xi \times \ln|\xi|$:

$$\Psi(t, \xi) = G(t, \xi)\sigma_{\ln}(0) + m_{\ln}(0),$$

where

$$G(t, \xi) = \frac{\text{sign}\xi \times \ln|\xi| - m_{\ln}(t)}{\sigma_{\ln}(t)},$$

and

$$\varphi_t(\xi) = \frac{1}{\sqrt{2\pi}\sigma_{\ln}(t)\xi} \exp\left\{ -\frac{[\text{sign}\xi \times \ln|\xi| - m_{\ln}(t)]^2}{2\sigma^2_{\ln}(t)} \right\}. \tag{7.10}$$

Hence, for processes of aging and wear that satisfy the condition $F = 0$, the instability is subject to the logarithmic normal distribution.

In practice, one often comes across values of F that are not equal to 1 or 0. For example, $F > 1$ is characteristic of processes with internal resistance to deterioration (elastic components, systems with feedback), and $F < 0$ for heterogeneous n-th order chemical reactions [52]. Transforming (7.8) to the form $[s^F(t)]' = (1/F)Q$ $[T, \alpha_1, ..., \alpha_n]$, we derive, by analogy with the preceding,

$$\mu(t, |\xi|^F) = \frac{1}{F}\left\{m'_F(t) + r_1(t)\left[|\xi|^F - m_F(t)\right]\right\},$$

where $r_1(t) = \sigma'_F(t)/\sigma_F(t)$, $m_F(t) = \int\limits_{-\infty}^{\infty} \text{sign}\xi|\xi|^F \varphi_t(\xi)\,d\xi$,

$$\sigma_F(t) = \left[\int\limits_{-\infty}^{\infty} \left[\text{sign}\xi|\xi|^F - m_F(t)\right]^2 \varphi_t(\xi)\,d\xi\right]^{0.5}.$$

Consequently, for the drift intensity in equations (7.1) and (7.3) is obtained

$$\mu(t, \xi) = \frac{1}{F}\text{sign}\xi(|\xi|)^{1-F}\left\{m'_F(t) + r_1(t)\left[\text{sign}\xi|\xi|^F - m_F(t)\right]\right\}. \tag{7.11}$$

From this, we derive for initial condition $\Psi(0, \xi) = \text{sign}\xi|\xi|^F$:

$\Psi(t, \xi) = G(t, \xi)\sigma_F(0) + m_F(0)$, where $G(t, \xi) = \frac{\text{sign}\xi|\xi|^F - m_F(t)}{\sigma_F(t)}$, and

$$\varphi_t(\xi) = \frac{|F|(|\xi|)^{F-1}}{\sqrt{2\pi}\sigma_F(t)} \exp\left\{-\frac{[\operatorname{sign}\xi(|\xi|)^F - m_F(t)]^2}{2\sigma_F^2(t)}\right\}. \tag{7.12}$$

This distribution, called the generalized normal distribution, is described in section 2.2.3. Analyzing this and the results obtained earlier in this section, one may conclude as follows:

the instability of an MI's MC is subject to the generalized normal distribution. Particular cases of this distribution are the normal distribution resulting when $F = 1$ and the logarithmic normal distribution when $F = 0$.

In principle, one may also substantiate other, more complex, modifications of the generalized normal distribution law if in equation (7.8) the factor $s^{1-F}(t)$ is replaced by a function $f[s(t)]$ of a different type (this is characteristic, for example, of processes of moistening material or of wear on the supports of measuring instruments [52]).

7.3   Criteria for Setting the Verification and Calibration Intervals

The following indicators of metrological reliability are used as criteria for setting VCIs:

- probability of metrological serviceability of an MI at a specific instant t: $P_{ms}(t)$;
- coefficient of metrological serviceability $K_{ms}(t)$,
- probability of failure-free operation over a specified time t: $P(t)$.

7.3.1   Probability of Metrological Serviceability

1. When an MI is being calibrated (or verified by a calibration method) the initial values of the MC are reestablished. Hence the error probability distribution of the MI over the set of MIs of one type after calibration has been done is equal to the initial density $\varphi_0(\xi)$, and after a run time (time) t after calibration is $\varphi_t(\xi)$, defined by formula (7.4). It follows from this that the probability of metrological serviceability at moment t for MIs subject to calibration is expressed by the formula

$$P_{Mu}(t) = P(t, -\Delta, \Delta) = \int_{-\Delta}^{\Delta} \varphi_t(\xi)\,d\xi = \int_{-\Delta}^{\Delta} \varphi_0[G(t, \xi)]\frac{\partial G(t, \xi)}{\partial \xi}\,d\xi$$

$$= \Phi[G(t, \Delta)] - \Phi[G(t, -\Delta)] = \Phi\left(\frac{\Delta^F - m_F(T)}{\sigma_F(T)}\right) - \Phi\left(\frac{-\Delta^F - m_F(T)}{\sigma_F(T)}\right),$$

$$\tag{7.13}$$

where $\Phi(x) = 1/\sqrt{2\pi} \int_{-\infty}^{x} e^{-0.5y^2}\, dy$ is the integral function of the normal distribution, $G(t, \xi)$ is the drift function of the MI's MC, Δ is the bound for allowable values of the MC.

2. During MI verification, the goal of which is to determine the fitness for use using the criterion of accuracy Δ, only those specimens whose error exceeds this norm are rejected and replaced by new or repaired units. Other specimens are not subject to calibration and leave the verification laboratory with the same error that they had when sent for verification. Hence the set of MIs sent to a verification laboratory at each instant will include specimens with a different run time since the last calibration: those that have been in operation one time interval t, two intervals $2t$, three intervals $3t$, and so forth. Evaluation of the metrological reliability of this set of MIs is possible within a model of the process of operation that has been set, characterized by the following conditions:

   - the duration of the process of using MIs of the given type (if specimens that have operated to the end of their term of service are replaced with new MIs) is not limited;
   - the average age of the set of MIs sent for verification, as well as the average indicators of accuracy and metrological reliability of this set, remain constant.

It was shown in [53] that for such a process to exist, it is necessary that, beginning at some instant, the number of new MIs entering this set be a constant quantity.

The maximum number n of verifications of one MI is $[T_{sl}/T]$, if T_{sl} is evenly divisible by T, and $[T_{sl}/T] + 1$ otherwise (T_{sl} is the service life of an MI, T is the VCI, $[x]$ is the integral part of x (the integral part of a number closest to it from below)). We find the probability $Q_{ms}(T)$ of metrological serviceability of the MIs sent for verification after VCI T and with a limited service life $T_{c\pi}$, using the formula for average probability:

$$Q_{ms}(T) = \sum_{s=1}^{n} Q_{\text{rej}}[(n-s)T]Q_s(T), \tag{7.14}$$

where $Q_{\text{rej}}[(n-s)T] = 1 - Q_{ms}[(n-s)T]$ is the probability of rejecting an MI and replacing it with a new one at the $(n-s)$-th verification after its launch into service, $Q_s(T) = \int_{-\Delta}^{\Delta} \varphi_{sT}(\xi)\, d\xi$ is the probability of metrological serviceability at the 1st, 2nd, …s-th verifications, including MIs that have a limited service life of T_{sl}. Substituting the function $\varphi_{sT}(\xi)$ into this formula, we obtain:

$$Q_s(T) = \max\{\Phi[B_s(T)] - \Phi[-A_s(T)],\ 0\},$$

where

$$A_1(T) = \frac{\Delta^F + m_F(T)}{\sigma_F(T)},$$

$$B_1(T) = \frac{\Delta^F - m_F(T)}{\sigma_F(T)},$$

and the other terms are recursively calculated by the formulas

$$A_i(T) = \min\left[\frac{\Delta^F + m_F(iT)}{\sigma_F(iT_i)}; A_{i-1}(T)\right],$$

$$B_i(T) = \min\left[\frac{\Delta^F - m_F(iT)}{\sigma_F(iT)}; B_{i-1}(T)\right], \quad i = 2, 3, ..., s.$$

During the established process of operating the set of MIs, $Q_{ms}[(n-s)T] = \text{const} = Q_{ms}(T)$. We substitute this function into (7.14):

$$Q_{ms}(T) = [1 - Q_{ms}(T)] \sum_{s=1}^{n} Q_s(T).$$

Transferring to the left side of the equation the terms containing $Q_{ms}(T)$, we derive

$$Q_{ms}(T) = \frac{\sum_{s=1}^{n} Q_s(T)}{1 + \sum_{s=1}^{n} Q_s(T)}.$$

On the other hand, as the unconditional probability of metrological serviceability of MIs with a limited service life of T_{sl}, it is equal to

$Q_{ms}(T) = P_{ms}(T)R(T_{sl})$, where $P_{ms}(T)$ is the conditional probability of the metrological serviceability of an MI under the condition that the run time of the MI is limited to T_{sl}, and $R(T_{sl})$ is the probability of this condition. Hence

$$P_{ms}(T) = \frac{1}{R(T_{sl})} \times \frac{\sum_{s=1}^{n} Q_s(T)}{1 + \sum_{s=1}^{n} Q_s(T)}. \tag{7.15}$$

The probability $R(T_{sl})$ is a constant that is the same for any values of $Q_s(T)$. Hence we set $Q_s(T) = 1$ for $s = 1, 2, ..., n$. Since in doing so there is no metrological failure in the set of MIs, then also $P_{ms}(T) = 1$. Substituting these values into (7.15), we derive:

$$1 = \frac{1}{R(T_{sl})} \times \frac{n}{1+n},$$

from which $R(T_{sl}) = n/(n+1)$. Hence, we finally have:

$$P_{ms}(T) = \frac{n+1}{n} \times \frac{\sum_{s=1}^{n} Q_s(T)}{1 + \sum_{s=1}^{n} Q_s(T)}, \qquad (7.16)$$

where

$$n = \begin{cases} \left[\frac{T_{sl}}{T}\right], & \text{if this number is an integer,} \\ \left[\frac{T_{sl}}{T}\right] + 1, & \text{if this number is not an integer.} \end{cases}$$

$$Q_s(T) = \max\{\Phi[B_s(T)] - \Phi[-A_s(T)],\ 0\},$$

$$A_1(T) = \frac{\Delta^F + m_F(T)}{\sigma_F(T)}, \qquad B_1(T) = \frac{\Delta^F - m_F(T)}{\sigma_F(T)},$$

and the remaining terms are recursively calculated by the formulas

$$A_i(T) = \min\left[\frac{\Delta^F + m_F(iT)}{\sigma_F(iT)};\ A_{i-1}(T)\right],$$

$$B_i(T) = \min\left[\frac{\Delta^F - m_F(iT)}{\sigma_F(iT)};\ B_{i-1}(T)\right], \qquad i = 2, 3, ..., s.$$

If the service life of the MIs is not limited, then we take $T_{c\pi} = \infty$. In this case

$$P_{ms}(T) = \frac{\sum_{s=1}^{\infty} Q_s(T)}{1 + \sum_{s=1}^{\infty} Q_s(T)}. \qquad (7.17)$$

7.3.2   Metrological Serviceability Coefficient

The metrological serviceability coefficient $K_{ms}(t)$ is equal to the mean percentage of time in the interval $(0,\ t)$ during which the MI is in a metrologically serviceable state. In accordance with this definition,, $K_{ms}(t)$ is computed using the formula

$$K_{ms}(T) = \frac{1}{T} \int_0^T P_{ms}(t)\, dt. \qquad (7.18)$$

7.3.3 Probability of Operation Without Metrological Failures (Probability of Failure-Free Operation)

Calculating the probability of failure-free operation is a much more complex task than predicting metrological reliability. It is usually estimated using the formula

$$
P(t) = \begin{cases} \dfrac{\Phi[G(t, \Delta)] - \Phi[G(t, -\Delta)]}{\Phi[G(0, \Delta)] - \Phi[G(0, -\Delta)]}, & \text{or} \\[2mm] \dfrac{\Phi[G(t, \Delta)]}{\Phi[G(0, \Delta)]} & \text{with the hypothesis of unilateral drift.} \end{cases}
\tag{7.19}
$$

However, this expression is not precise, since it estimates not the probability of failure-free operation in the interval $[0, T]$, but the probability of metrological serviceability of the MI at the terminal moment of this interval. Hence it can serve to estimate $P(t)$ only for those drift processes in which the trajectories are all monotonic functions. This is assumed by default, although most often is not especially referred to. The justification of such an approach is that the task of estimating the probability of outliers of non-stationary random processes beyond the specified limits (which in essence is the definition of $P(t)$) does not have an analytical solution in the general case.

At the same time, it turns out to be possible to find a solution to this problem for random processes with trajectories that are continuously differentiable with probability 1. A strict mathematical derivation of the unknown expression is rather complex.[3] Hence we shall limit ourselves to presenting an idea of the proof. It consists of the following.

We divide the interval of time $[0, T]$ into $n = 2^k$ $(k = 0, 1, 2, ...)$ equal parts equal to $\tau(k) = T/2^k$. Let us suppose that the metrological serviceability of an MI is checked at moments of time $t_i = i \times T/2^k$ $(i = 0.1, ..., n)$. The probability $P_{ms}(t_i)$ of metrological serviceability at each of the moments t_i is defined by (7.19). An occurrence consisting of the fact that the MI is metrologically serviceable at all these instants, taken together, is the intersection of the indicated occurrences[4]:

"metrological serviceability at $t_0, t_i, ..., t_n = \bigcap_{i=0}^{n} \{\text{metrological serviceability at } t_i\}$."

Consequently, the probability that the MI is metrologically serviceable at all these instants with the condition that it was metrologically serviceable at the initial moment, as the probability of the intersection of these occurrences, is equal to

---

[3] This derivation is presented in Chap. 9 of [47].
[4] The intersection of sets A and B is the set that is common to both A and B.

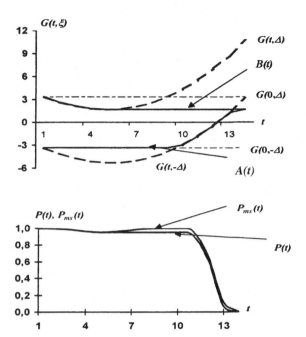

Fig. 25 Graphic illustration of curves to determine $P_{ms}(T)$ and $P(t)$. (**a**) Functions $G(t, -\Delta)$ and $G(t, \Delta)$ and their extreme values $A(t)$ and $B(t)$. (**b**) Graphs of $P_{ms}(T)$ and $P(t)$

$$P_{ms}(t_0, t_i, ..., t_n) = P\left[\bigcap_{i=0}^{n}\{metr.\,serv.\,at\,t_i/metr.\,serv.\,at\,t_0 = 0\}\right]$$
$$= \frac{\Phi[\min_{i=0,...,n} G(t_i, \Delta)] - \Phi[\max_{i=0,...,n} G(t_i, -\Delta)]}{\Phi[G(0, \Delta)] - \Phi[G(0, -\Delta)]}.$$

Now let us take k to infinity. Here, since $\tau(k) \to 0$, the points t_i will gradually fill the interval $[0, T]$, and if the passage to the limit is valid, then the probability of failure-free operation is defined as the limit as $k \to \infty$ of the probability of metrological serviceability of the MI at all 2^k points of this interval:

$$P(t) = \lim_{k=\infty} P_{ms}(0, t_1, ..., t_{2^k}) = \frac{\Phi[B(t)] - \Phi[A(t)]}{\Phi[G(0, \Delta)] - \Phi[G(0, -\Delta)]}, \qquad (7.20)$$

where $A(t) = \max_{\tau \in [0,t]} [G(\tau, -\Delta)]$, $B(t) = \min_{\tau \in [0,t]} [G(\tau, \Delta)]$.

Example 7.1. The limits of allowable systematic error of an MI are $\pm \Delta = \pm 1\%$. The initial error of the MI is determined by the calibration error and is characterized by $m(0) = 0$ and $\sigma(0) = 0.3\%$. The instability distribution of systematic error is subject to the normal law with $m(t) = (0.25t - 0.03t^2)$ %/month and $\sigma(t) = \sigma(0)$ $e^{0.01t}$%. $P_{ms}(t)$ and $P(t)$ are to be determined.

Figure 25a shows the curves of the functions $G(t,-\Delta)$, $G(t,\Delta)$ and their extreme $A(t)$ and $B(t)$. They reflect the non-monotonic nature of the drift of the systematic error of this type of MI. It is clear that in the interval 0–5 months, metrological failures occur only with intersection of the upper boundary of the region of allowable values, in the interval 10–15 months only with intersection of the lower boundary, and in the interval 5–10 months there were no metrological failures. The curves for $P_{ms}(t)$ and $P(t)$, shown in Fig. 25b, behave correspondingly. In particular, it is clear that $P_{ms}(t)$ increases in the interval 4–8 months, which is convincing evidence that it is not possible to estimate probability of failure-free operation using formula (7.19). The maximum difference between $P(t)$ and $P_{ms}(t)$ is 0.05 ($P(9\,\text{months.}) = 0.95$, while at the same time $P_{ms}(9\,\text{months}) = 1$).

7.4 Substantiation of Primary Verification and Calibration Intervals

When designating a primary VCI for new types of MIs released for use, the following types of sources of information on MI instability area possible: tests on the MIs or their separate units; data on instability of components of an MI; reliability indicators for the MI; data on the VCI for units that are analogous to the MI, as confirmed by experience using them.

The most preferable source, from the point of view of substantiating the VCI, is to test batches of the MIs for the purpose of evaluating their instability. These tests can be specially conducted (either in normal or in forced operational mode), and then combined with controlled reliability tests, or else introduced by monitored use of a setup batch. The procedure for conducting these tests, as regulated in [49, 50], is presented below.

A batch of MIs for conducting tests is formed. The volume N of the batch must be at least 30. The selected batch is subjected to tests in the usual or in accelerated mode (with a known acceleration factor). After several intervals of operating or running time Δt, the control parameters are measured.

The quantity Δt must be such that the increment $y_j(i\Delta t) = x_j(i\Delta t) - x_j[(i-1)\Delta t]$ in the MC can be measured with acceptable accuracy. This means that the $y_j(\Delta t)$ must be significant against the background of random measurement error. For the simplest, linear model of forecasting, the least squares method requires at least three groups of multiple measurements. Hence the duration of the tests must be at least $2\Delta t$.

Using the results of measurements of instability $y_j(i\Delta t)$, $j = 1, \ldots, N$ the sampling characteristics of the instability distribution of the MIs for the intervals $\Delta t, 2\Delta t, 3\Delta t, \ldots, n\Delta t$ are evaluated as follows.

1. Parameters F_i, $i = 1, 2, \ldots$ of the generalized normal distribution law for the MC $x_j(i\Delta t)$ (or of MC instability $y_j(i\Delta t)$) are found, to which information on the results of testing an MI of a given type corresponds, with the highest level of significance. The algorithm for statistical determination of this parameter:

   a. A matrix of values of F from 0 to +4 with step 0.1 is provided. For each value of this matrix and each $i\Delta t$ are found:

      - values $x_i^F = \operatorname{sign} x_j |x_j|^F$, $\quad j = 1, \ldots, N$;
      - sampling mean $m_F = 1/N \sum_{j=1}^N x_j^F$;
      - sampling standard deviation

      $$\sigma_F = \sqrt{\frac{\sum_{j=1}^N (x_j^F - m_F)^2}{N-1}};$$

      - value of the quantile of the χ^2- distribution

      $$\chi_F^2 = \sum_{j=1}^N \frac{(1 - 6\sigma_F \times f(x_j^F))^2}{6\sigma_F \times f(x_j^F)}.$$

   b. The value $\min_F \chi_F^2$ is found. For $F_i = F(i\Delta t)$ is taken the value that corresponds to this minimum.

      In accordance with the least squares method, this will be the best approximation of the sampling distribution of the MC (instability of the MCs) at instant $i\Delta t$ using the generalized normal distribution.

2. Sampling characteristics of the generalized normal distribution:

   - average value of parameter $F = 1/N \sum_{i=1}^N F_i$,
   - average value of $x^F - m_F(i\Delta t) = 1/N \sum_{j=1}^N \operatorname{sign}(x_j)(|x_j(i\Delta t)|)^F$, $i = 1, 2, \ldots$
   - standard deviation

   $$\sigma_F(i\Delta t) = \sqrt{1/(N-1) \sum_{i=1}^N [\operatorname{sign}(x_j)(|x_j(i\Delta t)|)^F - m_F(i\Delta t)]^2}$$

3. Using these values, approximating polynomials are selected by the least squares method for functions

   $$m_F(t) = m_a + m_b t, \quad \sigma_F^2(t) = \sigma_a^2 + \sigma_b^2 t. \tag{7.21}$$

4. The VCI is calculated using successive approximations from members of the series (0.25; 0.5; 1; 2; 3; 4; 5; 6; 7; 8; 9; 10; 11; 12; 15; 18; 21; 24; 30) months, and so forth, over 6 months. The order of calculation consists of the following:

   - the value of the VCI's T_1 is selected, equal to the actual value of the VCI's T;
   - the value of a criterion $R(T_1)$, established for each MI, is calculated in accordance with the formulas shown in Sect. 7.3 (for example, probability of metrological serviceability $P_{m5}(T_1)$);

- $R(T_1)$ is compared with the normalized value of criterion R^*. Metrological reliability of the MI is shown as higher than required if $R(T_1)>R^*$. In this case, the value of a VCI with $T_2>T_1$, closest to T_1 from above, is selected from members of the series shown in step 4. If the metrological reliability of the MI turns out to be lower than required $(R(T_1)<R^*)$, then the value $T_2<T_1$, closest to T_1 from below is selected;
- $R(T_2)$ is calculated, and then $R(T_2)$ and $R(T_1)$ are compared with R^*. If R^* is located between the values of $R(T_2)$ and $R(T_1)$, then the approximation is finished and the VCI is taken as $T = \min(T_1, T_2)$;
- if this condition is not satisfied, a value T_3 closest to T_2 is selected, and the operations indicated above are repeated. If it turns out that R^* is located between the values $R(T_2)$ and $R(T_3)$, then the VCI is taken as equal to $T = \min(T_2, T_3)$. Otherwise, T_4 closest to T_3, is selected, and so forth;
- the approximations continue until R^* ends up between $R(T_{i=1})$ and $R(T_i)$. Then $T = \min(T_{i-1}, T_i)$ is accepted.

7.5 Correction of Verification and Calibration Intervals in the Process of Operation of Measuring Equipment

The modes and conditions of operation (intensity of use, the predominant ranges of the measurand, frequency, and other non-information parameters of the input signal, measurements conditions, quality of technical and metrological mainte-nance, and so forth) of MIs of one type that are used on various objects can differ substantially from each other. In this regard, their metrological reliability can also diverge significantly, objective information about which is shown by the results of verifications and calibrations that are conducted. Hence the basic standards documents that regulate the manner of specifying the assignment of the VCI [49, 50] foresee the possibility of correcting the VCIs of particular specimens or groups of MIs during service, taking into account the results of their previous verifications and calibrations.

The correction procedure is defined in the form of information on the verification results.

7.5.1 Correction of Verification Intervals of a Group of MIs, with the Error Values of Each Specimen Recorded During Verification

First, the verification results are grouped by the index order of the verifications conducted after release of the MI from production or repair: the 1st group are the MIs that have had their first verification since manufacture or repair, the 2nd group are the MIs that have had their second verification since manufacture or repair, and so on. In each ith group there are N_i results of measuring the error of the MI (or its instability over the preceding VCI) x_j, $j = 1, ..., N_i$. If the results of all verifications are known, $N_i = N$, although generally $N_i \neq$ const.

Statistical analysis of the grouped verification results follows, and determination of the VCI according to the procedure presented in Sect. 7.3.

7.5.2 Correction of Verification Intervals of a Group of MIs with the Alternative Marker (Whether the MI Is Serviceable or Not), and the Error Sign of Each MI Recognized as Non-serviceable, Recorded During Verification

In this case, the boundedness of the input data does not permit obtaining an estimate of the VCI based on the generalized normal distribution. Hence the normal distribution of MI error is suggested. The calculations proceed as follows.

First, the verification results are grouped the same way as in Sect. 7.5.1.

Then statistical analysis of grouped verification results is conducted in the following manner.

1. Statistical probabilities are computed:

$$\bar{p}_{1i} = \bar{P}\{x(iT) \leq -\Delta\}, \quad \bar{p}_{2i} = \bar{P}\{x(iT) > \Delta\}, \tag{7.22}$$

where Δ is the limit of allowable MI error, and the quantiles of the normal distribution, $\bar{\lambda}_{1i}$ and $\bar{\lambda}_{2i}$, corresponding to them are defined by the equation $\bar{\lambda} = \Phi^{-1}(\bar{p})$, where $\Phi(\bar{\lambda}) = \frac{1}{\sqrt{2\pi}} \int_{-\infty}^{\bar{\lambda}} e^{-t^2/2} \, dt = \bar{p}$ is the standardized normal distribution function.

2. Statistical estimates of error are found: the average value $\bar{m}(iT)$ and standard deviation $\bar{\sigma}(iT)$

$$\bar{m}(iT) = \Delta \frac{\bar{\lambda}_{1i} + \bar{\lambda}_{2i}}{\bar{\lambda}_{1i} - \bar{\lambda}_{2i}}, \quad \bar{\sigma}(iT) = \frac{2\Delta}{\bar{\lambda}_{2i} - \bar{\lambda}_{1i}}, \, i = 1.2, \ldots \tag{7.23}$$

3. From the values of $\bar{m}(iT)$ and $\bar{\sigma}(iT)$, analogously to (7.21) an approximating polynomial is calculated for the functions $m(t)$ and $\sigma^2(t)$ of the normal distribution (where $F = 1$ is taken).

Thereafter, calculation of the VCI is done using the procedure set out in Sect. 7.4.

7.5.3   Correction of Verification Intervals of a Group of MIs with Just the Alternative Marker Regarding Serviceability of the MI Recorded During Verification

Likewise in this case, the normal distribution of MI error is suggested. Calculations proceeds as follows.

Verification results are grouped the same way as in Sect. 7.5.1.

Then the statistical probabilities $\bar{p}_i$ of recognizing the MI as serviceable are calculated from the results of the i-th verification.

A hypothesis is proposed regarding the "non-linear" random process of change of the MI error over time, characterized by the parameters

$$m(t) = 0, \ \sigma^2(t) = \sigma_a^2 + \sigma_b^2 t. \tag{7.24}$$

The first estimate T_1 of the VCI is calculated in the following manner.

1. Statistical estimates of the error drift parameters are found using this hypothesis

$$\overline{m}(iT) = 0, \quad \overline{\sigma}(iT) = \frac{\Delta}{\overline{\lambda}_{0.5(1+\bar{p}_i)}}, \ i = 1, 2 \ldots,$$

where $\overline{\lambda}_{0.5(1+\bar{p}_i)}$ is the quantile of the normal distribution corresponding to the probability $0.5(1 + \bar{p}_i)$.
2. From the values of $\overline{\sigma}(iT)$ an approximating polynomial (7.24) is calculated using the least squares method for the functions $\sigma^2(t)$.
3. The value T_1. of the VCI is calculated.

Then a hypothesis is formed regarding a linear random process of the change of MI error over time, characterized by the parameters

$$m(t) = m_a + m_b t, \quad \sigma(t) = \sigma_0. \tag{7.25}$$

A second estimate T_2 for the VCI is calculated as follows.

1. Statistical estimates of the error drift parameters are found using this hypothesis

$$\overline{m}(iT) = \Delta - \overline{\lambda}_{\bar{p}_i} \sigma_0, \quad \overline{\sigma}(iT) = \sigma_0,$$

where σ_0 is the SD of the calibration error of the MI (if this in unknown, it is taken as $\sigma_0 = \Delta/3$).
2. From the values of $\overline{m}(iT)$ an approximating polynomial (7.25) is calculated using the least squares method for the functions $m(t)$.
3. The value T2 for the VCI is determined.

Table 25 Sample estimates of parameters of model drift for the Sapfir 22DA measuring instrument

Index I of verification	Number n of serviceable MIs	$\bar{p}_i$	$0,5(1+\bar{p}_i)$	Non-linear random process $\bar{\lambda}_{0,5(1+\bar{p}_i)}$	$\bar{\sigma}(iT)$	$\bar{\sigma}^2(iT)$	Linear random process $\bar{\lambda}_{\bar{p}_i}$	$\bar{m}(iT)$
0	60	1	1	3	0.167	0.028	3	0
1	59	0.983	0.992	2.394	0.209	0.044	2.128	0.101
2	58	0.967	0.983	2.128	0.235	0.055	1.834	0.145
3	56	0.933	0.967	1.834	0.273	0.074	1.501	0.194
4	54	0.900	0.950	1.645	0.304	0.092	1.282	0.226
5	53	0.883	0.942	1.569	0.319	0.102	1.192	0.239

The lesser value of these derived estimates is taken as the VCI:

$$T = \min[T_1,\, T_2]. \tag{7.26}$$

Example 7.2. Verification of Sapfir 22DA instruments includes checking the correspondence of their error with the limit $\Delta = \pm0.5\%$ of allowable error. The service life of the instruments is $T_{sl} = 20$ years. Operation of 60 specimens of this type of MIs on an object with VCI at $T = 1.5$ years showed the following: one instrument each was rejected at the first, second, and fifth verifications, and two instruments each at the third and fourth verifications. We need to find the VCI corresponding to $P^*_{ms} = 0.95$.

First we estimate the statistical probabilities $\bar{p}_i$ of recognizing as serviceable the MIs at the ith verification with $\bar{p}_i = n_i/n_0$, where $n_0 = 60$ is the total number of MIs delivered for use, and n_i is the number of MIs recognized as serviceable at the ith verification. In this regard, the MIs brought into operation to replace the rejected ones were not considered. Then we find the quantiles $\bar{\lambda}_{0.5(1+\bar{p}_i)}$ and $\bar{\lambda}_{\bar{p}_i}$ and the parameters of the non-linear and linear drift models. Table 25 show the results of the calculations.

For the hypothesis of a non-linear random process, we assume $\bar{m}(iT) = 0$. For the hypothesis of a linear random process, we assume $\sigma(iT) = \Delta/3$.

Using the least squares method, we find the parameters of the drift models. For the linear model $m(t) = m_0 + m_1 t$, the system of normal equations has the form

$$\begin{cases} 5m_a + T\sum_{i=1}^{5} im_b = \sum_{i=1}^{5} \bar{m}(iT), \\ T\sum_{i=1}^{5} im_a + T^2\sum_{i=1}^{5} i^2 m_b = T\sum_{i=1}^{5} i\bar{m}(iT), \end{cases}$$

where $T = 1.5$ years is the effective VCI, $T\sum_{i=1}^{5} i = 1.5 \times 15 = 22.5$,

$$T^2\sum_{i=1}^{5} i^2 = 1.5^2 \times 55 = 123.75, \quad \sum_{i=1}^{5} \bar{m}(iT) = 0.905, \quad T\sum_{i=1}^{5} i\bar{m}(iT) = 4.606.$$

VCI T, years	$P_{ms1}(T)$	$P_{ms2}(T)$	min$\{P_{ms1}(T),$ $P_{ms2}(T)\}$
1.50	0.986	0.978	0.978
1.75	0.984	0.975	0.975
2.00	0.982	0.973	0.973
2.50	0.977	0.966	0.966
3.00	0.972	0.955	0.955
3.50	0.968	0.947	0.947

Table 26 Calculation of VCI for measuring instrument Sapfir 22DA

Hence, the system of normal equations has the form:

$$\begin{cases} 5m_a + 22.5m_b = 0.905, \\ 22.5m_a + 123.75m_b = 4.606. \end{cases}$$

Its solution is: $m_a = 0.074$ %, $m_b = 0.024$ %/year.

For the non-linear model $\sigma^2(t) = \sigma_a^2 + \sigma_b^2 t$, we have an analogous system of normal equations:

$$\begin{cases} 5\sigma_a^2 + 22.5\sigma_b^2 = 0.367, \\ 22.5\sigma_a^2 + 123.75\sigma_b^2 = 1.882. \end{cases}$$

Its solution is: $\sigma_a^2 = 0.0275$ (%)^2, $\sigma_b^2 = 0.0102$ (%)^2/year.

By formula (7.17) and aided by the MathCAD system, we find the probabilities of metrological serviceability of the MI for these two drift models. For the non-linear model,

$$m_1(iT) = 0, \ \sigma_1(iT) = \sqrt{0.0275 + 0.0102 \times i \times T},$$

$$Q_{s1}(T) = \Phi\left[\frac{0.5}{\sqrt{0.0275 + 0.0102 \times s \times T}}\right] - \Phi\left[-\frac{0.5}{\sqrt{0.0275 + 0.0102 \times s \times T}}\right].$$

For the linear model, $m_2(iT) = 0.074 + 0.024 \times i \times T$, $\sigma_2(iT) = \Delta/3 = 0.167$,

$$Q_{s2}(T) = \Phi\left[\frac{0.5 - (0.074 + 0.024 \times i \times T)}{0.167}\right] - \Phi[-3]$$

$$\cong \Phi[2.55 - 0.143 \times i \times T].$$

Thereafter, we select values T according to Sect. 7.4 and estimate with formula (7.16) the values of the probability of metrological serviceability from the non-linear drift model, $P_{ms1}(T)$, and from the linear model, $P_{ms2}(T)$. Table 26 shows the results of the calculations. They demonstrate that the optimal time for the VCI using the $P_{ms}^* = 0.95$ criterion is $T = 3$ years.

Chapter 8
Assurance of Measurement Accuracy in Compliance with ISO 5725 Standards

8.1 Basic Concepts

This chapter expounds upon the methodology of assurance of measurement accuracy, which is receiving ever greater dissemination in the country and in the world of metrology, and which is regulated by ISO 5725 international standards (which in Russia are adopted by state standards GOST R ISO 5725–1 through GOST R ISO 5725–6 [54]).

In accordance with the classical approach presented in Chap. 2, measurement accuracy is defined as the degree of closeness of a measurement result to the true (conventional) value of the measurand, wherein its value, determined with the aid of a primary standard, is taken as the conventional value of the quantity. But this approach is not applicable to many measurement tasks for which standards either are absent currently or cannot exist in principle. Such tasks include measurements of many elements of the chemical composition of substances and materials and objects of the natural environment; measurements of production quality indicators (such as in the petrochemical and food sectors of industry); and measurements in biology, medicine, and many others. In this case, the conventional value of a quantity in international and domestic practice is usually taken as the most probable value determined from the aggregated results of authoritative scientific or engineering studies. This value is called the accepted reference value of the quantity. It is defined as follows: *an accepted reference value is the value agreed upon for comparison and obtained as:*

1. *a theoretical or established value based on scientific principles;*
2. *an attributed or certified value based on experimental work of some national or international organization;*
3. *an agreed or certified value based on combined or experimental work under the guidance of a scientific or engineering group;*
4. *an expected value, i.e., the average value of a specified set of measurement results (but only if a), b), and c) are inaccessible).*

A.E. Fridman, *The Quality of Measurements: A Metrological Reference*,
DOI 10.1007/978-1-4614-1478-0_8, © Springer Science+Business Media, LLC 2012

This definition shows that a reference value of a key comparison, as studied in the preceding chapter, is formulated in complete compliance with the methodology of ISO 5725 standards.

As a result of this definition, the definition of *measurement accuracy* provided in Chapter 2 changes somewhat: it is understood as *the degree of proximity of measurement results to the true value (or if this is unavailable, to the accepted reference value)*. In turn, measurement accuracy results from two different properties of measurements: correctness and precision. *Correctness is the degree of proximity of the average value obtained, based on a large series of measurement results, to the accepted reference value.* This is characterized by *systematic error*, which is defined as *the difference between the expected value of the measurement results and the true value (or if this is unavailable, to the accepted reference value). Precision is the degree of proximity of independent measurement results to each other, obtained in specific regulated conditions.* It follows from this definition that precision has no relationship to the true or reference value of the measurand, but depends only on random measurement error.

The following statistical model is placed at the foundation of the methodology of the estimation of measurement accuracy. Measurements of some quantity are conducted in a group of measurement or test laboratories using the very same standardized method (for example, according to a standardized procedure of measurements (PM)). The result y_i of each measurement (here $i = 1, \ldots, n$ is the index number of the laboratory) represents the following sum:

$$y_i = m + B_i + e_i = \mu + \bar{\Delta} + B_i + e_i, \tag{8.1}$$

where $m = 1/n \sum_{i=1}^{n} y_i = \mu + \bar{\Delta}$ is the general average of the results of all measurements, which is the sum of the true (or accepted reference) value μ of the measurand and systematic error Δ_m, calculated on the set of all measurements of this value;

B_i is the systematic measurement error in the i-th laboratory due to the systematic errors of the MIs, deviation of measurement conditions from normal conditions, and other factors that depend on the technical state of the laboratory equipment and the staff's quality of work;

e_i is the random component of measurement error in the i-th laboratory.

The sum of systematic errors $\Delta_i = \Delta_m + B_i$ is the systematic error in the ith laboratory. Hence it is called *systematic error of the laboratory*. In accordance with the general definition of systematic error, the *systematic error of the laboratory (when executing a specific method) is the difference between the expected value of measurement results in a particular laboratory and the true (or if this is unavailable, the accepted reference) value of this quantity.*

The systematic error $\Delta_m = \bar{\Delta} = 1/n \sum_{i=1}^{n} \Delta_i$ is the same for all laboratories. It is caused by deficiencies of the selected measurements method that are common to all. Hence it is called systematic error of the measurement method. *By definition, systematic error of the measurement method is the difference between the expected value of the measurement results obtained at all laboratories that used this method, and the true (or if this is unavailable, the accepted reference) value of this quantity.*

By contrast, the value B_i of systematic error depends on the selection of laboratory. Hence it is called the laboratory component of systematic error of measurement. By definition, *the laboratory component of systematic error of measurement is the difference between the laboratory's systematic error when executing a specific method of measurement and the systematic error of the method of measurement.*

The spread of errors is estimated by the dispersion. The dispersion σ_w^2 of the random component e_i has the name *intra-laboratory dispersion.* It characterizes the spread of measurement results for the same conditions of measurement. These measurements are called *measurements in conditions of repeatability (conditions of repeatability are conditions in which independent measurement results are obtained by the same method, in the same laboratory, by the same operator, using the same equipment, within a short interval of time).* The dispersion σ_L^2 of systematic error B_i reflects the differences between laboratories. Hence it is called *inter-laboratory dispersion.* The dispersion of the sum $B_i + e_i$ of errors characterizes the spread of measurement results in all laboratories relative to the average measurement result m. Such measurements are called measurements in conditions of reproducibility (*conditions of reproducibility are conditions in which independent measurement results are obtained by the same method, in different laboratories, by different operators, using different equipment).*

Conditions of repeatability and reproducibility reflect two basic situations provided for in ISO 5725 standards. They define two different properties that characterize measurement precision: repeatability and reproducibility. *Repeatability is precision in conditions of repeatability, and reproducibility is precision in conditions of reproducibility.* Quantitatively, these properties are reflected in the standard (mean square) deviations and confidence intervals. *The standard deviation σ_r of repeatability is the standard deviation of results obtained in conditions of repeatability, and the standard deviation σ_R of reproducibility is the standard deviation of results obtained in conditions of reproducibility.* The limits are also defined analogously: *the repeatability limit r is the value that, with a confidence probability of 0.95, is not exceeded by the absolute value of the difference between two measurements results obtained in conditions of repeatability; the reproducibility limit R is the value that, with a confidence probability of 0.95, is not exceeded by the absolute value of the difference between two measurements results obtained in conditions of reproducibility.*

These indicators characterize measurement precision in the entire set of laboratories being examined. Hence they must be average estimations. These are introduced as follows. The standard deviations of repeatability and reproducibility are equal to the square root of their respective dispersions σ_r^2 and σ_R^2. The dispersion of repeatability is equal to the expected value of intra-laboratory dispersion in the set of all laboratories: $\sigma_r^2 = M(\sigma_w^2)$ (here M is the symbol for expected value). The dispersion of reproducibility is equal to the expected value of the dispersion that characterizes the spread of laboratory systematic errors, over the set of laboratories. Hence it is equal to the sum of the dispersion of repeatability and the inter-laboratory dispersion:

$$\sigma_R^2 = M(\sigma_w^2 + \sigma_L^2) = \sigma_r^2 + \sigma_L^2. \tag{8.2}$$

Table 27 Interrelationship of the basic concepts of ISO Series 5725 standards

Measurement properties	Accuracy		
		Precision	
	Correctness	Reproducibility	Repeatability
Measurement errors	Systematic error of the laboratory $\Delta_i = \Delta_m + B_i$		Random measurement error e_i
	Systematic error of the measurement method Δ_i	Laboratory component of systematic error B_i	
Estimates of measurement errors	$\Delta_m = \bar{\Delta} = \frac{1}{n} \sum_{i=1}^{n} \Delta_i$	Inter-laboratory dispersion σ_L^2	1. Dispersion of repeatability $\sigma_r^2 =$ intra-laboratory dispersion. 2. Standard deviation of repeatability σ_r. 3. Repeatability limit r.
		1. Dispersion of reproducibility $\sigma_R^2 = \sigma_r^2 + \sigma_L^2$. 2. Standard deviation of reproducibility σ_R. 3. Reproducibility limit R.	

From this follows the relationship between the standard deviations of repeatability and reproducibility:

$$\sigma_R = \sqrt{\sigma_r^2 + \sigma_L^2}. \tag{8.3}$$

Table 27 shows the framework for the concepts introduced in this section.

8.2   Basic Method for Estimation of the Measurement Method Precision

Experimental evaluation of the precision of a measurement method proceeds as follows. A set of measures, representing q different levels of the quantity (for example, a set of q specimens of the same mixture of substances), is sent to p laboratories. Each i-th laboratory performs n_{ij} measurements at each level j under conditions of repeatability. The following requirements are imposed on measurements conducted in the laboratories:

– each group of measurements must be done in observance of the conditions of repeatability;
– the mutual independence of the n_{ij} parallel measurements must be preserved;

- each series of the n_{ij} parallel measurements must be conducted in a short interval of time;
- measurements for all q levels of the quantity must be done by the same operator, and on the same equipment for each level;
- the time interval for performance of all measurements must be specified.

When planning the experiment, greater value is placed in the selection of the number n of parallel measurements and number p of laboratories, which define the scale of the experiment and the accuracy of the derived estimates. To substantiate the selection of these indicators, the extended uncertainty is estimated for the result of estimating the standard deviations σ_r of the repeatability and σ_R of the reproducibility. It is defined by the formula

$$P\{-A\sigma < s - \sigma < A\sigma\} = P, \tag{8.4}$$

where s is the standard deviation of the measurement results in the sample.

For a probability of $P = 0.95$, the following estimates of the deviation $s - \sigma$ are presented in the standard:

- for standard deviation of repeatability

$$A_r\sigma_r = 1.96\sigma_r\sqrt{\frac{1}{2p(n-1)}}, \tag{8.5}$$

- for standard deviation of reproducibility

$$A_R\sigma_R = 1.96\sigma_R\sqrt{\frac{p[1 + n(\gamma^2 - 1)] + (n-1)(p-1)}{2\gamma^4 n^2 p(p-1)}}, \tag{8.6}$$

where $\gamma = \sigma_R/\sigma_r$.

After constructing a database of the measurement results y_{ijk} ($k = 1, ..., n_{ij}$ is the index of the measurement) it is analyzed for the purpose of excluding outliers. The procedure for checking for outliers using the Grubbs criterion is presented in section 4.1. Then the average values of the measurement results in the base elements are determined (a base element is the combination of level of quantity and laboratory)

$$\bar{y}_{ij} = \frac{1}{n_{ij}} \sum_{k=1}^{n_{ij}} y_{ijk} \tag{8.7,}$$

and the standard deviation of the sample of a base element is

$$s_{ij} = \sqrt{\frac{1}{n_{ij} - 1} \sum_{k=1}^{n_{ij}} (y_{ijk} - \bar{y}_{ij})^2}. \tag{8.8}$$

If the base element contains only two measurement results, then it is taken as

$$s_{ij} = \frac{y_{ij1} - y_{ij2}}{\sqrt{2}}. \tag{8.9}$$

The sample estimates are then calculated:
- overall average value

$$\hat{m} = \bar{\bar{y}}_j = \frac{\sum_{i=1}^{p} n_{ij} \bar{y}_{ij}}{\sum_{i=1}^{p} n_{ij}}, \tag{8.10}$$

- dispersion of repeatability

$$s_{rj}^2 = \frac{\sum_{i=1}^{p} (n_{ij} - 1) s_{ij}^2}{\sum_{i=1}^{p} (n_{ij} - 1)}, \tag{8.11}$$

- inter-laboratory dispersions

$$s_{Lj}^2 = \frac{s_{dj}^2 - s_{rj}^2}{\bar{\bar{n}}_j}, \tag{8.12}$$

where

$$s_{dj}^2 = \frac{1}{p-1} \sum_{i=1}^{p} n_{ij} (\bar{y}_{ij} - \bar{\bar{y}}_j)^2, \tag{8.13}$$

$$\bar{\bar{n}}_j = \frac{1}{p-1} \left[\sum_{i=1}^{p} n_{ij} - \frac{\sum_{i=1}^{p} n_{ij}^2}{\sum_{i=1}^{p} n_{ij}} \right], \tag{8.14}$$

- dispersion of reproducibility

$$s_{Rj}^2 = s_{rj}^2 + s_{Lj}^2. \tag{8.15}$$

In the particular case when all $n_{ij} = n$, formulas (8.10-8.15) are simplified:

$$\hat{m} = \frac{1}{p} \sum_{i=1}^{p} \bar{y}_{ij}, \quad s_{rj}^2 = \frac{1}{p} \sum_{i=1}^{p} s_{ij}^2, \quad \bar{\bar{n}}_j = n, \quad s_{dj}^2 = \frac{n}{p-1} \sum_{i=1}^{p} (\bar{y}_{ij} - \bar{\bar{y}}_j)^2, \tag{8.16}$$

$$s_{Rj}^2 = \frac{1}{p-1} \sum_{i=1}^{p} (\bar{y}_{ij} - \bar{\bar{y}}_j)^2 + \left(1 - \frac{1}{n} \right) s_{rj}^2. \tag{8.17}$$

The values of the standard deviations σ_r of repeatability and σ_R of reproducibility are taken to be their sample estimates s_r and s_R.

8.3 Basic Method for Estimation of the Measurement Method Correctness

ISO Standard 5725–4 describes the basic methods of estimating measurement correctness. Estimation of measurement correctness is possible only when the accepted reference value μ of a quantity (see the error model at (8.1)) can be established experimentally as its conventional value, such as when using standards or standard specimens. These can be used as standard specimens:

- certified standard specimens,
- materials with known properties,
- materials whose properties were determined by an alternative method of measurement, for which its systematic error is known to be negligible.

In accordance with 8.1, the systematic error is a characteristic of measurement correctness. The standard treats two indicators of correctness: system error Δ_m of the measurement method and systematic error Δ_i of the laboratory.

8.3.1 Estimation of the Systematic Error of the Measurement Method

The purpose of the inter-laboratory test is not just to estimate the systematic error of the measurement method, but also to establish the fact of the statistical significance of this error. If it is established that the systematic error is not statistically significant, then it is necessary to find a maximum value for the systematic error for which it may remain undetected in this test.

The same requirements are presented for the tests as for estimating precision. The program for this test differs from the program for the test to estimate precision only by other criteria for the selection of the number p of laboratories and the number n of parallel measurements.[1] Since the extended uncertainty of the estimate of systematic error of the measurement method is equal to

$$1.96\sigma_R\sqrt{\frac{n(\gamma^2 - 1) + 1}{\gamma^2 pn}},$$

for a confidence probability of $P = 0.95$ for the results of the test, the minimum values of p and n must satisfy the inequality

$$1.96\sqrt{\frac{n(\gamma^2 - 1) + 1}{\gamma^2 pn}}\sigma_R \leq \frac{\Delta_M^*}{1.84}, \qquad (8.18)$$

---

[1] The standard treats the method of estimating systematic error of the measurements of a quantity at one level. Hence, the index j is dropped in the formulas presented.

where Δ_m^* is the specified (critical) value of the systematic error Δ_m of the measurement method at which it is considered significant.

Verification of precision, conducted at the first stage of an experiment, is accomplished by calculating the sample dispersion of repeatability s_r^2 and reproducibility s_R^2 from formulas (8.16–8.17), in which the index j is dropped and then these values are compared with values previously established and accepted for the true dispersions of repeatability σ_r^2 and reproducibility σ_R^2.

If σ_r was not defined previously, then s_r is considered to be its best estimate. Otherwise the statistic is defined:

$$C = \frac{s_r^2}{\sigma_r^2},\tag{8.19}$$

which is compared with the critical value $C_{\text{crit}} = \frac{\chi_{(1-\alpha)}^2(v)}{v}$ (and $\chi_{(1-\alpha)}^2(v)$ is understood to be the $(1-\alpha\%)$-th quantile[2] of the χ^2 distribution, with number of degrees of freedom $v = p(n-1)$). Usually $\alpha = 0.05$ is taken.

If $C \leq C_{\text{crit}}$, then it is not statistically significant if s_r^2 exceeds the value of the dispersion of repeatability σ_r^2. Hence the standard deviation of repeatability σ_r is used to estimate the systematic error of the method. If $C > C_{\text{crit}}$, then the occurrence $s_r^2 > \sigma_r^2$ is statistically significant. In this case it is essential to study the reasons for exceeding this value. As a result, it may turn out to be necessary to rerun the test.

Estimation of reproducibility proceeds analogously. If the standard deviation of reproducibility σ_R was not defined previously, then s_R will be considered to be its best estimate. If σ_R and σ_r were defined previously, the statistic is calculated:

$$C' = \frac{s_R^2 - ((n-1)/n)s_r^2}{\sigma_R^2 - ((n-1)/n)\sigma_r^2},\tag{8.20,}$$

which is compared with the critical value $C'_{\text{crit}} = \frac{\chi_{(1-\alpha)}^2(v)}{v}$. If $C' \leq C'_{\text{crit}}$, the numerator of (8.20) exceeding its denominator is statistically insignificant. Here, σ_R and σ_r can be used to estimate measurement correctness. If $C' > C'_{\text{crit}}$, this exceedance is statistically significant. In this case, a careful study of the conditions of measurements in each laboratory, for the purpose of finding the reasons for the divergence, must be conducted before estimating the systematic error of the method of measurement. As a result, it may turn out to be necessary to rerun the test.

The estimate of the systematic error of the measurement method is found by the formula

$$\hat{\Delta}_M = \hat{m} - \mu,\tag{8.21}$$

where $\hat{m} = \bar{\bar{y}}$ is calculated by formula (8.10),
μ is the accepted reference value of the quantity.

---

[2] The value of a centered (with mean 0) and normalized (with standard deviation 1) random quantity, corresponding to the value of the $(1 - \alpha)$ distribution function

The uncertainty of this estimate is expressed by the standard deviation

$$\sigma_{\hat{\delta}} = \begin{cases} \sqrt{\frac{\sigma_R^2 - (1-1/n)\sigma_r^2}{p}}, & \text{if } \sigma_R, \ \sigma_r \text{ are unknown,} \\ \sqrt{\frac{s_R^2 - (1-1/n)s_r^2}{p}}, & \text{if they are unknown.} \end{cases} \tag{8.22}$$

The extended uncertainty of the systematic error Δ_m for a probability coverage of 0.95 can be approximately calculated by the formula

$$\hat{\Delta}_M - A\sigma_R \leq \Delta_M \leq \hat{\Delta}_M + A\sigma_R, \tag{8.23}$$

where $A = 1.96\sqrt{\frac{n(\gamma^2-1)+1}{\gamma^2 pn}}$.

If this interval includes the value of zero, the systemic error is statistically insignificant at the $\alpha = 0.05$ significance level. Otherwise, a conclusion is drawn as to the existence of systematic error with average value $\hat{\Delta}_m$.

8.3.2   Estimation of the Systematic Error of the Laboratory

An intra-laboratory test, performed for the purpose of estimating the systematic error Δ_i of the laboratory, must strictly correspond to the standard for measurement procedure, and the measurements must be performed in conditions of repeatability. Before an estimation of correctness, it is necessary to verify measurement precision by comparing the intra-laboratory standard deviation with the established standard deviation σ_r of the method. The test program includes measurements performed in one laboratory in a test to estimate precision. An additional requirement is the use of an accepted reference value μ.

The n of measurement results must satisfy the inequality

$$\frac{1.96\sigma_r}{\sqrt{n}} \leq \frac{\Delta_i^*}{1.84}, \tag{8.24}$$

where Δ_i^* is the specified (critical) value of the systematic error of the laboratory at which it is considered significant.

In analyzing measurements results, the mean $\bar{y}_w$ and standard deviation s_w are first calculated by formulas (7) and (8), in which the indices i and j are dropped. The measurement results must be studied for the presence of outliers using the Grubbs criterion (see section 4.1). Then the statistic is calculated

$$C'' = \frac{s_w^2}{\sigma_r^2} \tag{8.25}$$

and compared with the critical value $C''_{crit} = \chi^2_{(1-\alpha)}(v)/v$, where $v = n - 1$. If $C'' \leq C''_{crit}$, then the estimate s_w^2 exceeding the value of the dispersion of repeatability σ_r^2 is statistically insignificant. Hence the standard deviation of repeatability σ_r is used to estimate systematic error of the laboratory. If $C'' > C''_{crit}$, then the occurrence $s_w^2 > \sigma_r^2$ is statistically significant. In this case, it is necessary to study the question of repeating the test, with confirmation at all stages that the measurement method is being executed by the appropriate method.

Then the systematic error of the laboratory is calculated using the formula

$$\hat{\Delta}_i = \bar{y}_w - \mu. \tag{8.26}$$

The uncertainty of this estimate is expressed by the standard deviation

$$\sigma_{\hat{\delta}} = \begin{cases} \frac{\sigma_r}{\sqrt{n}}, & \text{if } \sigma_r \text{ is known,} \\ \frac{s_r}{\sqrt{n}}, & \text{if it is not known.} \end{cases} \tag{8.27}$$

The extended uncertainty of the systematic error of the laboratory Δ_i with coverage probability 0.95 is calculated by the formula

$$\hat{\Delta}_i - A_w\sigma_r \leq \Delta_i \leq \hat{\Delta}_i + A_w\sigma_r, \tag{8.28}$$

where

$$A_w = \frac{1.96}{\sqrt{n}}. \tag{8.29}$$

If this interval includes the zero value, the systematic error of the laboratory is statistically insignificant at the $\alpha = 0.05$ significance level. Otherwise, it must be considered significant.

8.4   Application of the Repeatability and Reproducibility Limits

In accordance with 8.1, the repeatability limit r is the value that, with confidence probability of 0.95, is not exceeded by the absolute value of the difference $|y_{ij} - y_{ik}|$ of two measurement results y_{ij}, y_{ik} obtained in conditions of repeatability. Analogously, the reproducibility limit R is the value that, with confidence probability of 0.95, is not exceeded by the absolute value of the difference $|y_{ij} - y_{kl}|$ of two measurement results y_{ij}, y_{kl} obtained in conditions of reproducibility. Since the dispersion of the difference of two random values is the sum of their dispersions, the standard deviation of the difference between two measurements in conditions of repeatability is equal to $\sigma = \sqrt{2}\sigma_r$, where σ_r is the standard deviation of repeatability. For a normal distribution, by which random measurement errors are generally governed, the quantile $\lambda = 1.96$ corresponds to a confidence probability of 0.95. Hence the repeatability limit is equal to

$$r = \lambda\sigma = 1.96 \times \sqrt{2}\sigma_r = 2.77\sigma_r \cong 2.8\sigma_r. \tag{8.30}$$

Analogously, the reproducibility limit is $R = 2.8\sigma_R$. Consequently, the critical (i.e., maximum) value of the difference $|y_1 - y_2|$ between two results of measurements performed in one laboratory is equal to

$$CD = r = 2.8\sigma_r, \tag{8.31}$$

and in different laboratories is

$$CD = R = 2.8\sigma_R. \tag{8.32}$$

If two groups of measurements are performed in conditions of repeatability: n_1 measurements with average value $\bar{y}_1$ and n_2 measurements with average value $\bar{y}_2$, then the dispersion of the difference $(\bar{y}_1 - \bar{y}_2)$ is

$$\sigma^2 = \sigma_r^2 \left(\frac{1}{n_1} + \frac{1}{n_2} \right).$$

Consequently, the critical difference of the two results of multiple measurements $|\bar{y}_1 - \bar{y}_2|$ in conditions of repeatability will be equal to

$$CD = 1.96 \times \sqrt{2}\sigma_r \sqrt{\frac{1}{2n_1} + \frac{1}{2n_2}} = 2.8\sigma_r \sqrt{\frac{1}{2n_1} + \frac{1}{2n_2}} = r\sqrt{\frac{1}{2n_1} + \frac{1}{2n_2}}. \tag{8.33}$$

Substituting $n_1 = n_2 = 1$ into this formula, we derive expression (8.30).

Current quality control systems for goods and services involve constant monitoring of the stability of the production process (including monitoring accuracy of measurements) for the purpose of timely detection of moments of degradation of its quality and elimination of the reasons for such degradation. Monitoring the stability is based on function (8.33): if for two successive measurements of the monitored quality indicator the results $\bar{y}_i$ and $\bar{y}_{i+1}$ from multiple measurements (with measurement indices n_i and n_{i+1}) satisfy the condition

$$|\bar{y}_{i+1} - \bar{y}_i| \le r\sqrt{\frac{1}{2n_{i+1}} + \frac{1}{2n_i}}, \tag{8.34}$$

then this divergence between them may be caused by random measurement errors. Hence it is presumed that the quality indicator being monitored has not changed in value. If condition (8.34) is not fulfilled, then it is necessary to do a new adjustment of the equipment or else take other measures to correct the production process.

If groups of measurements of the same quantity are done in different laboratories, and in each laboratory under conditions of repeatability, then the dispersion of the difference $\bar{y}_1 - \bar{y}_2$ will be equal to

$$\sigma^2 = \left(\sigma_L^2 + \frac{\sigma_r^2}{n_1}\right) + \left(\sigma_L^2 + \frac{\sigma_r^2}{n_2}\right) = 2\sigma_L^2 + \sigma_r^2\left(\frac{1}{n_1} + \frac{1}{n_2}\right)$$

$$= 2(\sigma_L^2 + \sigma_r^2) - 2\sigma_r^2\left(1 - \frac{1}{2n_1} - \frac{1}{2n_2}\right).$$

Considering formula (8.30) and the identity $\sigma_L^2 + \sigma_r^2 = \sigma_R^2$, we derive, for the critical difference of the two results of multiple measurements $|\bar{y}_1 - \bar{y}_2|$ in conditions of reproducibility,

$$CD = 1.96 \times \sqrt{2} \times \sqrt{\sigma_R^2 - \sigma_r^2\left(1 - \frac{1}{2n_1} - \frac{1}{2n_2}\right)}$$

$$= \sqrt{R^2 - r^2\left(1 - \frac{1}{2n_1} - \frac{1}{2n_2}\right)}. \tag{8.35}$$

It is evident that formula (8.31) is a particular case of this formula for $n_1 = n_2 = 1$.

These functions are used in practice in the resolution of disputes between suppliers and buyers regarding the quality of delivered production. If the result $\bar{y}_1$ of the supplier's measuring a quality indicator y and the result $\bar{y}_2$ of its measurement by the buyer satisfy the inequality

$$|\bar{y}_2 - \bar{y}_1| \leq \sqrt{R^2 - r^2\left(1 - \frac{1}{2n_1} - \frac{1}{2n_2}\right)}, \tag{8.36}$$

then these measurement results do not contradict each other. In this case, their average $y = (\bar{y}_1 + \bar{y}_2)/2$ is taken as the quality indicator, and the calculation regarding production proceeds using this value. If condition (8.36) is not fulfilled, then that provides evidence of systematic measurement error by the supplier or (and) buyer of production. In this case it is usual to bring in an arbitration laboratory to resolve the dispute. The arbitration laboratory will conduct a measurement of the quality indicator, the result of which, $\bar{y}_3$, is compared with $\bar{y}_1$ and $\bar{y}_2$. If, for example, the pair of values $\bar{y}_3$ and $\bar{y}_1$ satisfy condition (8.36), but $\bar{y}_3$ and $\bar{y}_2$ do not satisfy it, then the average $y = (\bar{y}_1 + \bar{y}_3)/2$ is used for the conventional value y of the indicator.

If an accepted reference value μ_0 of the quantity is known, it is possible to estimate the systematic measurement error. Its definition consists of comparing the result $\bar{y}$ of multiple measurements with the accepted reference value μ_0 of the quantity, which by definition has no error. Hence, if the systematic error of the laboratory is estimated as $(\bar{y} - \mu_0)$, then its dispersion is equal to

$$\sigma^2(\bar{y}) = \sigma_L^2 + \frac{\sigma_r^2}{n} = \frac{2(\sigma_L^2 + \sigma_r^2) - 2\sigma_r^2(1 - (1/h))}{2} = \frac{1}{2}\left(2\sigma_R^2 - 2\sigma_r^2\frac{n-1}{n}\right).$$

Consequently, the critical difference for this estimate is equal to

$$CD = 1.96 \times \frac{1}{\sqrt{2}} \sqrt{2\sigma_R^2 - 2\sigma_r^2 \frac{n-1}{n}} = \frac{1}{\sqrt{2}} \sqrt{R^2 - \frac{n-1}{n} r^2}. \qquad (8.37)$$

If this condition is satisfied:

$$|\bar{y} - \mu_o| \le \frac{1}{\sqrt{2}} \sqrt{R^2 - \frac{n-1}{n} r^2}, \qquad (8.38)$$

then this means that the systematic measurement error is not significant and one may presume that there is none. If (8.38) is not satisfied, then this provides evidence that the difference $(\bar{y} - \mu_0)$ is statistically significant. Then this value is taken as the estimate of the systematic error of the laboratory.

Precisely the same is estimating the systematic error of the method. In this case, the overall average $\bar{\bar{y}} = 1/p \sum_{i=1}^{p} \bar{y}_i$, obtained at p laboratories, is compared with the reference value (here $\bar{y}_i = 1/n_i \sum_{j=1}^{n_i} y_{ij}$ is the result of multiple measurements of the quantity at the ith laboratory). The dispersion of the value of $\bar{\bar{y}}$ is

$$\sigma^2(\bar{\bar{y}}) = \frac{1}{p^2} \sum_{i=1}^{p} \sigma^2(\bar{y}_i) = \frac{1}{p^2} \left(p\sigma_L^2 + \sigma_r^2 \sum_{i=1}^{p} \sum_{j=1}^{n_i} \frac{1}{n_i^2} \right) = \frac{1}{p} \sigma_L^2 + \frac{1}{p^2} \sigma_r^2 \sum_{i=1}^{p} \frac{1}{n_i}$$

$$= \frac{1}{2p} \left[2(\sigma_L^2 + \sigma_r^2) - 2\sigma_r^2 + \frac{2\sigma_r^2}{p} \sum_{i=1}^{p} \frac{1}{n_i} \right] = \frac{1}{2p} \left[2\sigma_R^2 - 2\sigma_r^2 \left(1 - \frac{1}{p} \sum_{i=1}^{p} \frac{1}{n_i} \right) \right].$$

Consequently, the critical difference of the value $\bar{\bar{y}} - \mu_o$ is equal to

$$CD = 1.96\sigma(\bar{\bar{y}}) = \frac{1}{\sqrt{2p}} \sqrt{R^2 - r^2 \left(1 - \frac{1}{p} \sum_{i=1}^{p} \frac{1}{n_i} \right)}. \qquad (8.39)$$

Further estimation proceeds in a manner analogous to the estimation of the systematic error of a laboratory.

8.5   Method for Testing the Acceptability of Measurement Results

Current quality systems in calibration and testing laboratories prescribe the use of various methods to increase the reliability of measurement information that is obtained. One of these methods is to conduct repeated measurements and compare the obtained results. In doing so, the measurement results are considered acceptable if they do not contradict each other, since the deviation between them is caused only by random measurement error. The preceding section presented the calculated functions for determining the allowable limits of these deviations, the *CD*s or critical differences. This section presents typical procedures used in such verification, and rules for determining the final measurement result.

Let us suppose that in one laboratory with conditions of repeatability, two results y_1 and y_2 of single measurements of one value are obtained. If this condition is satisfied:

$$|y_1 - y_2| \leq r, \tag{8.40}$$

then both results are recognized as acceptable and the final measurement result is indicated as their average: $\bar{y} = (y_1 + y_2)/2$. If condition (8.40) is not satisfied, then the procedure is as follows.

1. If it does not cost much to obtain results, then two more measurements y_3 and y_4 are completed. The range $D(4) = (y_{max} - y_{min})$ of the sample is found from the 4 measurement results found. Then from Table 24, the value $f(4) = 3.6$ of the coefficient of critical range[3] $f(n)$ is found for number of measurements $n = 4$.

   The measurements results are considered acceptable if this condition is satisfied:

$$D(n) \leq CR_{0.95}(n), \tag{8.41}$$

which suggests that the deviations between them are caused by random measurement error, which is subject to the normal distribution with zero mean and dispersion σ_r^2. Consequently, if $D(4) \leq 3.6\sigma_r$, the results of the 4 measurements are considered acceptable. In this case, their average $\bar{y} = (y_1 + y_2 + y_3 + y_4)/4$ is taken as the final measurement result. If this condition is not satisfied, the measurements results are not consonant with this hypothesis. In this case, the median $Me = (y_{(2)} + y_{(3)})/2$ is taken as the final measurement result ($y_{(2)}$ and $y_{(3)}$ are the second and third results of this ordered series), since for distribution laws different from the normal law, the median is the best estimate (Table 28).

   The critical range values for a confidence probability of 0.95 and sample size n is defined by the formula

$$CR_{0.95}(n) = f(n)\sigma_r. \tag{8.42}$$

Hence the critical range value is $CR_{0.95}(4) = 3.6\sigma_r$.

2. If an obtained measurement is costly (for example, it takes several days, or it is involved with using expensive equipment and expenditures for costly standard specimens and reagents), just one more measurement y_3 is completed, and the range $D(3) = (y_{max} - y_{min})$ of the sample from the 3 measurements results

---

[3] *The coefficient of critical range $f(n)$ is the 95% quantile of the distribution of the range of the sample and values subject to the normal distribution with mean equal to 0 and standard deviation equal to 1.*

Table 28 Critical range coefficients $f(n)$

n	$f(n)$	n	$f(n)$	n	$f(n)$
2	2.8	17	4.9	32	5.3
3	3.3	18	4.9	33	5.4
4	3.6	19	5.0	34	5.4
5	3.9	20	5.0	35	5.4
6	4.0	21	5.0	36	5.4
7	4.2	22	5.1	37	5.4
8	4.3	23	5.1	38	5.5
9	4.4	24	5.1	39	5.5
10	4.5	25	5.2	40	5.5
11	4.6	26	5.2	45	5.6
12	4.6	27	5.2	50	5.6
13	4.7	28	5.3	60	5.8
14	4.7	29	5.3	70	5.9
15	4.8	30	5.3	85	6.0
16	4.8	31	5.3	100	6.1

obtained is found. Then the value $f(3) = 3.3$ and the critical range value $CR_{0.95}(3) = 3.3\sigma_r$ is found from Table 28. If $D(3) \leq 3.3\sigma_r$, the results of the 3 measurements are considered acceptable, and their average value $\bar{y} = (y_1 + y_2 + y_3)/3$ is taken as the final measurements result. For $D(3) > 3.3\sigma_r$ and it is not possible to conduct yet another measurement, the median $Me = y_{(2)}$ is taken as the final measurements result (here $y_{(2)}$ is the second result in the ordered series); if there is a possibility of conducting one more measurement, then the final measurements result is obtained in accordance with option 1.

Example 8.1. Determining the Content of Gold in a Copper Concentrate Using an Assay [54].

The standard deviation of repeatability is $\sigma_r = 0.12\,\text{g/t}$, and the repeatability limit is $r = 2.8 \times 0.12 = 0.34\,\text{g/t}$. Measurements results $y_1 = 11.0\,\text{g/t}$, $y_2 = 10.5\,\text{g/t}$ were obtained. Since $|y_1 - y_2| = 11.0 - 10.5 = 05\,\text{g/t} > r = 0.34\,\text{g/t}$, a third measurement is conducted. The result $y_3 = 11.0\,\text{g/t}$ is obtained. Since $D(3) = (y_{\max} - y_{\min}) = 0.5\,\text{g/t} > CR_{0.95}(3) = 3.3 \times 0.12 = 0.40\text{g/t}$, the measurement results are considered unacceptable. In connection with the fact that it is possible to conduct one more measurement, the result $y_4 = 10.8\,\text{g/t}$ was obtained. In the order series (10.5; 10.8; 11.0; 11.0) g/t, $y_{\max} = 11.0\,\text{g/t}$, $y_{\min} = 10.5\text{g/t}$, average $\bar{y} = (10.5 + 10.8 + 11.0 + 11.0)/4 = 10.8\text{g/t}$, and median $Me = (10.8 + 11.0)/2 = 10.9\text{g/t}$. Since $D(4) = (y_{\max} - y_{\min}) = 0.5\text{g/t} > CR_{0.95}(4) = 3.6 \times 0.12 = 0.43\text{g/t}$, the measurement results are considered unacceptable. Hence $Me = 10.9\text{g/t}$ is taken as the final measurement result.

8.6 Applicability of the Repeatability Indices During Approval of Measurement Procedures

One of the manifestations of scientific and technical progress is the development of new measurement techniques based on physical principles not used earlier for this purpose. As a result, new measurement procedures are created, that are more complete by comparison with widely used methods that are regulated by measurement procedures (MPs) that were confirmed (standardized) earlier. With this in mind, in order that the introduction of a new method into practice not disturb the unity of this type of measurements, it is essential to make its scale correspond with the scale of measurements of the standardized MP. In other words, to introduce a new measurement method into practice, one must transfer to it the unit size or scale being used in this type of measurement. If the measurements of this type of are relative, i.e., implemented with the aid of MIs that are calibrated in the specified unit, the task is easily resolved – by including the MI that performs the developed measurement method into the system of transferring the size of this unit. But if, as often happens in analytical measurements, no such system has been constructed or there are generally no MIs, but the measurements are done by performing procedures described in an MP, then the simplest and soundest method of transferring a unit size or scale to the new measurement method is to certify it using the standardized MP. This certification is done as follows.

1. In one laboratory, n measurements are done in conditions of repeatability, in accordance with the certifiable and standardized MP at several points of the range of measurements. Usually these points are taken at 20%, 50%, and 80% of the upper limit of measurements. Two series of results are obtained for the n-fold measurements:

$$\bar{x}_1 = \frac{1}{n} \sum_{j=1}^{n} x_{1j}, \dots, \bar{x}_l = \frac{1}{n} \sum_{j=1}^{n} x_{lj}$$

   for the certifiable MP, and analogously $\bar{x}_{01}, \dots, \bar{x}_{0l}$ for the standardized MP.

2. The differences $\Delta_i = \bar{x}_i - \bar{x}_{0i}$ in the measurement results are calculated and their significance in comparison to random errors of the certifiable and standardized MPs is analyzed. The indicators of repeatability are used for this purpose.

   If $|\Delta_i| \leq \sqrt{\frac{r^2 + r_0^2}{2n}}$, where r, r_0 are the limits of repeatability of the certifiable and standardized MPS, then the systematic deviations in the measurement results obtained by the comparative methods are insignificant, and one may bypass considering them inestimating the uncertainty of the certifiable MP.

   If $|\Delta_i| > \sqrt{\frac{r^2 + r_0^2}{2n}}$, then these deviations are statistically significant. In this case, the systematic error of the certifiable MPO is evaluated. For this purpose, the dependence of the values of the absolute measurement error Δ_i or relative error $\delta_i = \Delta_i/x_{0i}$ on the measurand is analyzed. If one of these dependencies is approximately constant, then an appropriate correction $\Delta x = \sum_{i=1}^{l} \Delta_i / l$ or $\delta x = \sum_{i=1}^{l} \delta_i / l$ is introduced along the entire range of measurements.

If a monotonic change Δ_i or δ_i is noticed in the range of measurements, then this correction is determined using a linear or piecewise-linear approximation. These corrections are introduced to the results of measurements conducted using the certifiable MP. The corrected differences of measurement results are equal to $\tilde{\Delta}_i = \bar{x}_i - \bar{x}_{0i} - \Delta x$ (or $\tilde{\Delta}_i = \bar{x}_i - \bar{x}_{0i}(1 + \delta x)$).

3. The type A standard uncertainties of the differences $\tilde{\Delta}_i$ are estimated:

$$u_A = \frac{1}{l-1} \sum_{i=1}^{l} \left(\tilde{\Delta}_i - \tilde{\Delta} \right)^2, \tag{8.43}$$

where $\tilde{\Delta} = 1/k \sum_{i=1}^{k} \tilde{\Delta}_i$.

4. The extended uncertainty of the measurement result on the certifiable MP is estimated using the formula

$$U = ku, \tag{8.44}$$

where $u = \sqrt{u_A^2 + u_0^2}$ is the standard uncertainty of the measurement result on the certifiable MP,

$u_0 = U_0 / k_0$ is the standard uncertainty of the measurement result on the standardized MP,

U_0 and k_0 is the extended uncertainty of the measurement result on the standardized MP and its coverage factor,

k is the coverage factor of the extended uncertainty of the measurement result on the certifiable MP with coverage probability 0.95.

References

1. Malikov M.F. Fundamentals of Metrology. M., Kommertspribor, 1949, 477 pp. (In Russian.)
2. Burdun G.D., Markov B.N. Fundamentals of Metrology. M., Izdatelstvo Standartov, 1972. (In Russian.)
3. Tyurin N.I. Introduction to Metrology. M., Izdatelstvo Standartov, 1973, 279 pp. (In Russian.)
4. Selivanov M.N., Fridman A.E., Kudryashova Zh.F. Quality of Measurement. Metrological Reference Book. L., Lenizdat. 1987, 295 pp. (In Russian.)
5. Shishkin I.F. Theoretical Metrology - M.: Izdatelstvo Standartov, 1991, 471 pp. (In Russian.)
6. Kozlov M.G. Metrology. SPb., MGUP Publishing House "Mir Knigi", 1998, 107 pp. (In Russian.)
7. Sergeyev A.G., Krokhin V.V. Metrology. Tutorial. M., Logos, 2000, 408 pp. (In Russian.)
8. Bryansky L.N., Doynikov A.S., Krupin B.N. Metrology. Scales, Measurement Standards, Practice. M., VNIIFTRI, 2004, 222 pp. (In Russian.)
9. GOST 16263-70. State System of Measurement Assurance. Metrology. Terms and Definitions. M., Izdatelstvo Standartov, 1972, 52 pp. (In Russian.)
10. Kamke D., Kremer K. Physical Grounds of Measurement Units. M., Mir, 1980, 208 pp. (In Russian.)
11. International Vocabulary of Metrology – Basic and General Concepts and Associated Terms VIM, 3 ed., Final 2007-05-18, Joint Committee for Guides in Metrology CIPM, 2007, 146 p.
12. Gorelik D.O., Konopelko L.A., Pankov E.D. Ecological Monitoring. Optoelectronic Instruments and Systems. T.2, SPb, "Krismas+",1998, 592 pp. (In Russian.)
13. Gauss K.F. Selected Geophysical Works, translated from Latin and German. T. 1, M., 1957, pp. 89–109. (In Russian.)
14. Fridlender I.G. The Law of Distribution of Random Measurement Errors. Works of the Zaporozhye Institute of Agricultural Machinery Industry, issue 2, K., Mashgiz, 1955, pp. 13–20. (In Russian.)
15. Fridman A.E. Generalized normal distribution law of errors. Metrologia, 39, 2002, 241–247 p.
16. Kramer G. Mathematical Statistics Methods. M., Mir, 1975, 648 pp. (In Russian.)
17. GOST 8.009-84. State System of Measurement Assurance. Normalizable Metrological Characteristics of Measuring Equipment. (In Russian.)
18. RD-453-84. Procedural Instructions. Characteristics of the Errors of Measuring Equipment in the Actual Operating Conditions. Calculation Methods. (In Russian.)
19. Guide to the Expression of Uncertainty in Measurement. 2nd ed., Geneva, ISO, 1995, 101 p.
20. EUROCHEM/SITAK Guide "Quantitative Description of Uncertainty in Analytical Measurements", 2nd ed. / translated from English under the editorship of L.A. Konopelko.- SPb, D.I. Mendeleyev Institute for Metrology (VNIIM), 2002. - 141 pp. (In Russian.)

A.E. Fridman, *The Quality of Measurements: A Metrological Reference*,
DOI 10.1007/978-1-4614-1478-0, © Springer Science+Business Media, LLC 2012

21. GOST 8.207-76. State System of Measurement Assurance. Direct Measurements with Repeated Observations. Methods for Processing the Observation Results. Fundamentals. (In Russian.)

22. GOST 8.381-80. State System of Measurement Assurance. Measurement Standards. Ways of Expression of Uncertainties. (In Russian.)

23. Publication Reference EA-4/02. Expression of the Uncertainty of Measurement in Calibration. European Co-operation for Accreditation, 1999, 79 p.

24. Golubev E.A., Isayev L.K. Measurement. Control. Quality. GOST R 5725. Fundamentals. Matters of Implementation and Application /-M.: Standardinform, 2005. 136 pp. (In Russian.)

25. GOST R ISO 5725–2. Accuracy (Correctness and Precision) of the Methods and Measurement Results. Part 2. The Basic Method for Determination of Repeatability and Reproducibility of the Standard Measuring Method. (In Russian.)

26. Fridman A.E. New Methodology of Processing the Multiple Measurement Results. Izmeritelnaya Tekhnika, 2001, No. 11, pp.54–59. (In Russian.)

27. International Standard ISO 31–2000. Quantities and Units.

28. GOST 8.417-02. State System of Measurement Assurance. Units of Quantities. (In Russian.)

29. Recommendation No. 1 (CI-2005). Preparative steps towards new definitions of the kilogram, the Ampere, the Kelvin and the mole in terms of fundamental constants. Recommendations of the International committee for weights and measures.

30. Sena L.A. Units of Physical Quantities and Their Dimensions. M., Nauka, 1977, 336 pp. (In Russian.)

31. Stotsky L.R. Physical Quantities and Their Units. Reference Book. M., Prosvescheniye, 1984, 239 pp. (In Russian.)

32. Aleksandrov Yu.I. Points at Issue of Modern Metrology in Chemical Analysis /- SPb, Publishing House of the B.E. Vedeneyev VNIIG, 2003, 304 pp. (In Russian.)

33. Institute of the Russian Language of the Academy of Science of the USSR. Dictionary of the Russian Language, v.1. Publishing House "Russky Yazyk", M., 1981, 696 pp. (In Russian.)

34. RMG 29–99. State System of Measurement Assurance. Metrology. Basic Terms and Definitions. Minsk, Izdatelstvo Standartov, 2000, 47 pp. (In Russian.)

35. GOST R 8.558-08. State System of Measurement Assurance. State Measurement Chain for Temperature Measuring Instruments. (In Russian.)

36. CIPM. National and International Needs in the Field of Metrology: International Cooperation and Role of the BIPM. Preprint to the Meeting of the NMI Directors, 23–25 February, 1998.

37. GOST 8.061-07. State System of Measurement Assurance. Content and Formation of Measurement Chains. (In Russian.)

38. GOST 8.578-08. State System of Measurement Assurance. State Measurement Chain for Instrument Measuring the Content of Components in Gas Mixtures. (In Russian.)

39. Okrepilov V.V. Quality Control. SPb, Nauka, 2000, 911 pp. (In Russian.)

40. The World of Metrology at the Service of the Mankind. The BIPM Director A.J. Wallard's Message to the World Day of Metrology, 20 May, 2006.

41. CIPM. Mutual Recognition Arrangement (MRA). CIPM. Mutual Recognition of the National Measurement Standard and Calibration and Measurement Certificates Issued by the National Metrology Institutes (NMI). Arrangement drawn by the BIPM and signed by the NMI Directors.

42. Guidelines for CIPM key comparisons. 1 March, 1999. With modifications by the CIPM in October 2003, 9 p.

43. Chunovkina. A.G., Kharitonov I.A. Evaluation of Regional Key Comparison Data: Two Approaches for Data Processing, Metrologia, 43, 2006.

44. COOMET R/GM/14:2006. Guide on Estimation of the COOMET Key Comparison Data. (In Russian.)

45. htpp://kcdb.bipm.org

46. Fridman A.E. Theory of Metrological Reliability of Measuring Equipment. Izmeritelnaya Tekhnika, 1991, No. 11, pp.3–11. (In Russian.)

47. Fridman A.E. Theory of Metrological Reliability of Measuring Equipment. Composite mono-graph "Fundamental Problems of the Theory of Accuracy", SPb, Nauka, 2001, pp. 382–413. (In Russian.)
48. Fridman A.E. Metrological Reliability of Measuring Equipment and Determination of Verifi-cation Intervals. Metrologia, 1991, No. 9, pp. 52–61.
49. GOST 8.565-99. State System of Measurement Assurance. The Order for Establishment and Correction of Verification Intervals of Measurement Standards. (In Russian.)
50. RMG 74–2004. Guide on Intergovernmental Standardization. State System of Measurement Assurance. Methods for Determination of Verification and Calibration Intervals of Measuring Instruments. (In Russian.)
51. Fridman A.E. Estimation of Metrological Reliability of Measuring Instruments and Multivalue Measures. Izmeritelnaya Tekhnika, 1993, No. 5, pp.7–10. (In Russian.)
52. Reference Book "Reliability and Effectiveness in Technical Equipment ", v. 10, M., Publish-ing House "Mashinostroyeniye", 1990, pp. 254–257. (In Russian.)
53. Fridman A.E. The Link between the Reliability and Accuracy Indices of the Population of Measuring Instruments. Works of Metrology Institutes of the USSR, issue 200 (260) "General Matters of Metrology", Leningrad, Energiya, 1977, pp. 51–60. (In Russian.)
54. GOST R ISO 5725-1 – GOST R ISO 5725-6. Accuracy (Correctness and Precision) of the Measuring Methods and Measurement Results. (In Russian.)

Index

Regression
  differential equation, 161, 162
  properties, 162
Relative measurement, 9–10
Relative measurement error, 24–25
Relative measurement indeterminacy, 58
Reliability, 157, 159–167, 170, 172, 174, 191
Repeatability
  definition, 181
  dispersion, 184, 186
  standard deviation, 181, 183, 193
Repeatability indices, applicability of, 194–195
Repeatability limit, 181
  application, 188–191
  standard deviation, 193
Reproducibility
  definition, 181
  dispersion, 184, 186
  estimation, 186
  standard deviation, 181, 183
Reproducibility limit, 181, 188–191
Reproducing distribution, 33
Residual bias (RB), 26
Residuals, 97–100, 102, 103, 107
RF Ministry of Industry and Energy, 146
RMO Consultative Committees, 150
RMOs. *See* Regional metrological
  organizations (RMOs)
ROFBs. *See* Radio-frequency optical bridges
  (ROFBs)
Rostekhregulirovaniye. *See* Federal Agency for
  Technical Regulation and Metrology

S
Salinity measurements, 3
Sampling units/sample preparation units, 17
Sapfir 22DA instruments, verification of, 177–178
Scale and time transformation, 16
Scale transformation (linear/non-linear), 16
Scientific Research Institute for Standard
  Reference Data (NIISSD), 19
SD. *See* Standard deviation (SD)
Secondary standards, 126–128
Second, definition of, 114
Second-order statistical error, 136, 137
Second-order working standards, 128
Secretary of State for Trade and Industry, 145
Sena, L.A., 121
Sensor, 16
Sets of measures, 18
Sharply divergent results/outliers, 73
Shirokov, K.P., 6, 124

SI. *See* International system of units (SI)
Single measurement, 7–8
Single-valued measures, 18
Skewness, 36, 38
Special standards, 128
SSRS. *See* State standard reference
  sample (SSRS)
Stability, of measuring instrument, 159
Standard deviation (SD), 30, 36, 39, 47, 56, 61,
  63, 74, 75, 81, 88, 131, 137, 140, 160,
  161, 163, 164, 173, 175, 187, 188
  of base element, 183
  error of measurement result, 59
  MI, instability of, 159
  of repeatability, 181–184, 186, 188, 193
  repeatability limit, 193
  of reproducibility, 181–184, 186
  uncertainty, 188
Standardization, of MPs, 15, 46, 148, 149, 194
Standardized metrological properties, 15
Standard reference data category, 19
Standards, 8, 9, 15, 21, 62, 111–112, 114, 116,
  120, 121, 124, 134, 135, 140, 141,
  143–155, 174, 179–195
Standard samples, 18, 20
Standards base, 126
Standards installation, 21
Standards of units, 126–132
Standard uncertainty, 56, 61–62, 79, 85,
  90–92, 109, 137
  of conditional equation(s), 100, 102, 103, 107
  definition, 60
  four-terminal resistor, electrical
    resistance of, 69
  of input quantities, 68
  MC, instability of, 160
  of measurement result, 59, 94, 195
  of measurements, 141
  of unit size, 140, 141
State measurement chain, 141–145
State metrology control, 146
State primary standard, of unit of
  temperature, 130
State standard reference sample (SSRS), 79–80
State standards, 131
Static measurement, 8, 12
Statistical distribution, of random
  measurement error, 38–41
  χ^2 distribution, 38–39
  fisher's distribution, 41
  student's distribution, 39–41
Statistical errors, 135
Statistical valuations, theory of, 35